The Connections World

A central feature of modern Asia that trumps differences in economic and political systems is the web of close relationships running between and within business and politics; *the connections world.* These networks facilitate highly transactional interactions yielding significant reciprocal benefits. Although the connections world has not as yet seriously impeded Asia's economic renaissance, it comes with significant costs and fallibilities. These include the creation and entrenchment of huge market power and the attenuation of competition. They in turn hold back the growth in productivity and innovation that will be essential for further development. The connections world also breeds massive inequalities that may culminate in political instability. The authors argue that if Asia's claim to the 21st century is not to be derailed, major changes must be made to policy and behaviour so as to cut away the foundations of the connections world and promote more sustainable economic and political systems.

SIMON COMMANDER is Managing Partner of Altura Partners and Visiting Professor of Economics at IE Business School in Madrid.

SAUL ESTRIN is Professor Emeritus of Managerial Economics and Strategy at London School of Economics.

The Connections World

The Future of Asian Capitalism

SIMON COMMANDER
Altura Partners, London

SAUL ESTRIN
London School of Economics and Political Science

CAMBRIDGE
UNIVERSITY PRESS

University Printing House, Cambridge CB2 8BS, United Kingdom

One Liberty Plaza, 20th Floor, New York, NY 10006, USA

477 Williamstown Road, Port Melbourne, VIC 3207, Australia

314–321, 3rd Floor, Plot 3, Splendor Forum, Jasola District Centre,
New Delhi – 110025, India

103 Penang Road, #05-06/07, Visioncrest Commercial, Singapore 238467

Cambridge University Press is part of the University of Cambridge.

It furthers the University's mission by disseminating knowledge in the pursuit of education,
learning, and research at the highest international levels of excellence.

www.cambridge.org
Information on this title: www.cambridge.org/9781009169790
DOI: 10.1017/9781009169783

© Simon Commander and Saul Estrin 2022

First published 2022

A catalogue record for this publication is available from the British Library.

Library of Congress Cataloging-in-Publication Data
Names: Commander, Simon, author. | Estrin, Saul, author.
Title: The connections world : the future of Asian capitalism / Simon Commander, IE Business
   School, Madrid, Saul Estrin, London School of Economics and Political Science.
Description: 1 Edition. | New York, NY : Cambridge University Press, [2022] |
   Includes bibliographical references and index.
Identifiers: LCCN 2021063125 (print) | LCCN 2021063126 (ebook) | ISBN 9781009169790
   (hardback) | ISBN 9781009169776 (paperback) | ISBN 9781009169783 (epub)
Subjects: LCSH: Business networks–Asia. | Asia–Economic conditions. | Capitalism–Asia. |
   BISAC: BUSINESS & ECONOMICS / Development / Economic Development
Classification: LCC HD69.S8 C66157 2022 (print) | LCC HD69.S8 (ebook) |
   DDC 384.55/4095–dc23/eng/20220228
LC record available at https://lccn.loc.gov/2021063125
LC ebook record available at https://lccn.loc.gov/2021063126

ISBN 978-1-009-16979-0 Hardback
ISBN 978-1-009-16977-6 Paperback

SC: In memory of Amar and John

SE: For Jenny

Contents

Colour Plates section to be found between pp. 77 and 78

Figures

Colour Plates

Tables

Acknowledgements

This book was initially conceived in the spring of 2019 whilst one of us was in India and the other in Indonesia. It evolved slowly but surely through many discussions and scribblings on boards once we were both back in London, before settling in to writing around the time that the pandemic began to rear its head. We had been friends and, for a time, colleagues over several decades. Those proved to be good foundations. Writing our book has been an immensely enjoyable and harmonious experience; a strong affirmation of friendship, common interest and pleasure in exploring ideas.

Subsequently, Sumon Bhaumik, Michael Carney, Jeni Klugman, John Llewellyn and George Magnus have all read and commented in detail on a draft of the book as did three anonymous reviewers. Without implicating them, their comments have helped improve it substantially.

Over the years and during the writing, we have had many discussions with colleagues and friends about the themes covered in the book. We would like to thank David Audretsch, Maurizio Bussolo, Stephan Chambers, Ajay Chhibber, Maryann Feldman, Ravi Kanbur, John Llewellyn, George Magnus, Costas Markides, Branko Milanović, the late Mario Nuti, Donald Peck, Stavros Poupakis, Daniel Shapiro and Sean Wilentz.

The book has relied to a very substantial extent on wonderful research support from some very talented and hard-working young people. Aside from combing endless data resources, Rebeca Granda Marcos and Thamashi de Silva have been invaluable by helping us frame and refine our arguments, not least by commenting in detail on all the chapters. Yuan Hu was very helpful in the analysis of market structure in Chapter 4. Christopher Commander provided very useful research assistance for Chapters 3 and 4.

Thanks to both the London Library and the LSE Library and their staff for access to the marvellous resources. At Cambridge University Press, it has been a pleasure working with our editor, Phil Good. The Research Infrastructure and Investment Fund of LSE's Department of Management provided generous support, and Nino Nizharadze and Emma Ward were both exceptionally helpful. Ellie Cumpsty carefully read through the final manuscript, Rachel Harris put it into shape and Ruth Martin compiled the index.

Finally, it is common for authors to thank their long-suffering families for their forbearance. In our case, the space that the pandemic created for us to concentrate on writing, rather than making a nuisance of ourselves, was doubtless an unintended blessing! Even so, special thanks to Jenny, Zarela, Christopher and Olivia.

Abbreviations

1MDB	1Malaysia Development Berhad
ADB	Asian Development Bank
AI	artificial intelligence
ANDAs	abbreviated new drug applications
APEC	Asia-Pacific Economic Cooperation
B2C	business-to-consumer
BJP	Bharatiya Janata Party
BPO	business process outsourcing
BRI	Belt and Road Initiative
CCP	Chinese Communist Party
CI	confidence interval
COVID-19	coronavirus disease 2019
CR	concentration ratio
CSR	corporate social responsibility
DMFs	drug master files
EDGAR	Emission Database for Global Atmospheric Research
EIU	Economist Intelligence Unit
ESCAP	Economic and Social Commission for Asia and the Pacific
EU	European Union
EVs	electric vehicles
FDI	foreign direct investment
FISIM	financial intermediary services indirectly measured
GDP	gross domestic product
GEM	Global Entrepreneurship Monitor
GVCs	global value chains
ICT	information and communication technologies
ILO	International Labour Organization
IMF	International Monetary Fund

IPO	initial public offering
IT	information technology
M&A	mergers and acquisitions
MNEs	multinational enterprises
NASSCOM	National Association of Software and Service Companies
NCE	new chemical entity
NGO	non-governmental organization
OECD	Organisation for Economic Co-operation and Development
PCC	Philippine Competition Commission
PISA	Programme for International Student Assessment
PPP	purchasing power parity
R&D	research and development
SASAC	The State-owned Assets Supervision and Administration Commission of the State Council
SEZs	special economic zones
SMEs	small- and medium-sized enterprises
SOEs	state-owned enterprises
TEA	Total early-stage Entrepreneurial Activity
TFP	total factor productivity
UNCTAD	United Nations Conference on Trade and Development
UNDP	United Nations Development Programme
UNIDO	United Nations Industrial Development Organization
VAT	value-added tax
VC	venture capital
VIE	variable interest entity
WDI	World Development Indicators
WGI	Worldwide Governance Indicators
WIPI	Wireless Internet Platform for Interoperability
WTO	World Trade Organization

I The Strengths and Fallibilities of Asian Capitalism

I.I PERCEPTIONS OF ASIA

Asia is vast and varied, its physical contours subject to many different demarcations. For many centuries, European chroniclers considered that Asia started at Constantinople, although over time this boundary was displaced eastwards. In this book, the Asia that we are talking about starts thousands of miles east of the Bosphorus in the flat and densely populated Indus and Gangetic plains of Pakistan and India before traversing the massive and desolate highland spaces of the Himalayas and Tibet, passing into the very diverse topologies of the many Chinese provinces and at its eastern perimeter, the Koreas, before falling into the sea opposite Japan. Beneath China lie the states of Southeast Asia, stretching from Myanmar and Thailand through to the Mekong Basin with Vietnam curled around its outer edge, while further south stretch the elongated archipelagos of Malaysia and then Indonesia, the latter extending far in the direction of the Antipodes. Over this immense terrain, it is scarcely any wonder that disparities in climate, ecology, social and political organisation and culture are so large. Yet in recent decades, there has been a marked tendency to speak as much about regional attributes as those at a national or local level. Indeed, talk of an Asian miracle or the Asian twenty-first century has become a new staple.

Until Vasco da Gama's voyage to India in 1497–99, European cartography and knowledge had extended no further than western Persia and the Gulf, despite the chronicles of some earlier explorers. Thereafter, as the frontiers of territory and knowledge were pushed back and gradually revealed, later explorers and visitors were often dazzled by the splendour of Asia's courts and rulers but also the quality of its products, natural and man-made. For example, China's abilities in science and technology were comparatively advanced.[1] In 1700, India alone accounted for around a quarter of the world economy and a similar share of the global textile trade. The Chinese economy was only slightly smaller in that year.[2]

Yet, as innovation and growth picked up in western Europe in the late eighteenth and nineteenth centuries, the desire to dominate trade elided into a desire to dominate territory and with it came the colonial moment in which most of Asia fell under the direct sway of one or more European power. Despite acute rivalry between those powers and a recognition of the political and strategic differences across dominions, intellectual currents, such as orientalism – a fashion for pooling traits of behaviour and systems of rule – often simply rolled up most of the Asiatic world into a common space, albeit one with attributes that were deemed mostly outmoded, if not reprehensible, and almost always inferior to their European comparators. But even when devoid of colonial condescension, it has not been uncommon for more recent writers to portray Asian government as inherently different from its more western counterparts. For instance, historian Karl Wittfogel argued in the 1950s that the Orient was doomed to despotism due to the pre-eminent need to harness and allocate water resources through the implementation of large-scale public infrastructure works. He also argued that this induced a profound continuity so that, for example, communist rule in China was in many respects similar to earlier systems.[3]

Although such ways of typifying the world have by no means entirely disappeared, more modern narratives about Asia tread a rather different path, balanced once again between marvel at its recent and dramatic successes – not least the massive cumulative growth in

---

[1] A point established at length in Joseph Needham's (1954–2015) magisterial volumes on Chinese science and technology.

[2] Maddison (2010); see also www.visualcapitalist.com/2000-years-economic-history-one-chart/.

[3] Wittfogel (1957).

income achieved since 1980 – but also the dangers and threats that this resurrection poses to the dominant world order. Those dangers may come from the participation of giant countries such as India and China in global trade and production and the ensuing consequences for workers in the advanced economies of Europe, Japan and North America, but also from the accretion of political and military power that has accompanied economic success. China's growing nationalist rhetoric and expansive claims to territory and influence in the region have proven, unsurprisingly, to be unsettling. More generally, China's extraordinary growth in the size of its economy and in the average income of its people has also unleashed dire prophecies of future dominance and threats to American hegemony, in particular.[4]

Whatever the inferences and particular interpretations, it is quite clear that Asia's re-emergence as a grand regional and, increasingly, global force now focusses interest in ways that could scarcely have been imagined even fifty or so years ago. Then, the dominant narrative was to bemoan the vast amount of entrenched poverty, especially in the Indian subcontinent, as well as the periodic excesses of communist rule, such as the Great Leap Forward and the appalling famine that ensued, in China. Now, it is more about whether Asia's resurgence will result in a region that rivals either North America or Europe. This rivalry extends way beyond the political to embrace technological and productive capacity, including the ability to innovate.

To begin to address these questions presupposes, of course, that the direction of travel that has been unleashed this past half-century will be sustained and that the foundations of greater prosperity that have been laid prove to be exactly that. Here, there is no single voice among the myriad number of commentators, whether in relation to the future of the region as a whole or at the level of individual countries. Some have suggested that these economies will struggle to attain rich country levels of income because of institutional and

---

[4] For example, Spalding (2019).

other failings that will hold them back. One consequence is likely to be the inability to create large knowledge-intensive sectors of economic activity that are innovative. This notion is sometimes summarised as the middle-income trap. Others point to the ability of some of these countries not only to marshal resources and to create new sectors and activities, but the way that this has been in innovative spaces, such as software and artificial intelligence (AI), that many would have expected to be the domain of the richer world. But whatever the balance of interpretations, politicians and citizens in the region have increasingly adopted a more optimistic tone – including through responses to public opinion surveys – about their futures and the respective places of their economies in the global system, especially in the two giant countries, China and India. Recognition of this weight and dynamic has also been reflected in secondary ways, such as the composition of the G20 or voting rights in international organisations. Perhaps most significantly, it is clear that attempts to address carbon emissions and climate change cannot succeed without Asian action both on the ground and in terms of accepting constraints within the context of international agreements.

I.2 PILLARS OF ASIA'S RESURGENCE: COMMONALITIES TRUMP PARTICULARITIES

Given Asia's extraordinary renaissance and the resulting recalibration of the world economy, our concern in this book is with understanding whether that resurgence can be expected to retain its vitality so that these countries can continue along a path towards substantially greater wealth and opportunity. To do that requires, of course, that we understand very well how Asia has got to the position that it currently finds itself in and what have become the main characteristics of these economies following decades of rapid expansion. We should also clarify that when talking about Asia in this book, we are primarily concerned with the larger emerging economies of the region. These are indicated in Figure 1.1, where the countries that form the focus of this book are named. Although Japan and the smaller

Map of Asia

Source: Natural Earth, 1.50m Cultural Vectors: Admin 0-Countries

FIGURE I.I Map of Asia

island states – Taiwan and Singapore – are not our main focus, their various experiences post-1945 in the former case and since the 1970s for the latter are, of course, very relevant in understanding the policy models followed by the countries on which we are concentrating. Indeed, it is very clear that South Korea – an economy that has successfully become a high-income economy – based much of its strategy on Japanese post–Second World War experience. And, in turn, China has aimed to pursue policies that have been tried and tested in South Korea. In short, these earlier experiences of development are used to cast light on what has happened – and what is likely to happen – in those economies of Asia that are our main interest.

What is evident is that the various models that have driven Asia's transformation have been strikingly different from earlier templates of capitalist development that accompanied the ascent of the older rich worlds of Europe and North America. Moreover, despite some major differences within Asia – China and Vietnam (not to forget the current-day anomaly that is North Korea) pursued a Soviet-like path of the planned economy for several decades, something that was not followed in most other countries – there are some surprisingly powerful and common features that cut across differences in political systems, institutional organisation and geography and that also intrude substantially into underlying patterns of economic behaviour and governance.

These common features are centred on the pervasive use of connections – familial, commercial and political. They trump the particularities of countries' political systems and their associated institutions. Even when countries have made transitions from autocracy to democracy, it is striking how such networks of connections have survived and entrenched themselves. We term this resilient and powerful phenomenon as the *connections world*. It is in no small measure due to the way in which such connections between businesses, politicians and the state have played out that Asia has been able to achieve so much cumulative growth. But this path comes with its costs, in terms of both how these societies and economies currently function and their potential functioning in the future.

A notable feature of the various Asian capitalisms has been the pivotal role of the state and public policy in driving growth and productivity. The intellectual and practical precursor was post-1945 Japan.[5] Using state guidance of the economy through industrial policy while mobilising public resources to stimulate selected sectors and activities was an approach that was then explicitly imitated by South Korea, Singapore and Taiwan, among others. Most other countries in the region also relied on a prominent role for the state, along with an active industrial policy, but in ways that generally involved protection of domestic industries and rarely involved the successful nurturing of new activities and sources of productivity growth. In China and

---

[5] But also, elements of Meiji Japan; see Beasley (2018).

Vietnam, the entire economy effectively became subject to the priorities established by government. Not surprisingly, this led to massive inefficiencies but also some remarkable accumulations of productive capacity and knowledge in activities that benefitted from support by government and the related channelling of resources. In almost all instances, the leading role of the state included the establishment of major state-owned enterprises (SOEs) across wide swathes of the respective economies.

The role of an activist state has been widely acknowledged as a driving factor behind Asia's success. What has been far less widely discussed is the way in which the state and the private sector have interacted and engaged with each other. To our mind, as striking – and undoubtedly more long lasting – has been the way in which many Asian economies have come to be populated by often substantial, highly influential and acquisitive private businesses, many of whom have been, and continue to be, organised in family-owned and -controlled business groups. Business groups are confederations of firms that are bound together in both formal and informal ways, including in many instances through ownership vested in families and dynasties. Ownership and control are often highly opaque, and many business groups suffer from weak governance and oversight. Such entities have tended to benefit from public largesse or preferential access to assets, finance and other sorts of privilege, including of a regulatory variety. Consequently, not only have the boundaries between public and private been difficult to draw but major pockets of private market power and economic concentration – sometimes explicitly fashioned by the actions of the state – have also been created. With this have come networks of economic and political influence that web together politicians, political organisations and business. Such networks have proven very capable of perpetuating themselves even while tolerating some changes in composition and shape. As we shall see later, the consequences of these organisational forms and the networks that underpin them have by no means been unambiguously adverse, but they have often had deleterious effects at both economic and political levels. Those costs have proven difficult to address, not least because their network nature has made them far more able to resist attempts at change.

1.3   CONNECTING BUSINESS AND POLITICS

The connections between politics and business take many forms, and these forms depend in part on the political and institutional arrangements that exist in each country. Even so, there are several, recurring patterns that emerge irrespective of these institutional and other differences, such as the following. One-time public officials commonly choose to move directly into political life, either standing for office or taking up non-elected appointments with clear political dimensions. Similarly, businesspeople very often choose to move explicitly into the political sphere – a wide selection of prime ministers and presidents, such as Nawaz Sharif in Pakistan, the Rajapaksa family in Sri Lanka and a fair number of recent Filipino presidents – are highly visible cases in point. The process also proceeds in a very widespread way at lower levels of the political hierarchy, including at provincial and municipal levels. Although the motives vary significantly, a common motivation is the perceived need to protect, or further, their business interests. In a similar vein, private businesses tend to make financial and other donations to political parties or campaigns – sometimes within the legal limits but often outside those limits. Public officials or politicians may also be shareholders – sometimes openly but more often covertly – in private businesses.

The ties between politics and business materialise in a large and diverse set of ways. Figure 1.2 lists the channels of interaction that run between them. For example, among the more common manifestations are the awarding of public contracts to favoured businesses or business groups by politicians. This may or may not occur with side payments or bribes, but almost always there is some underlying reciprocity or bargain involved. Well-connected companies may also be able to garner access to finance in amounts that may not be warranted or on terms that can be preferential, such as through

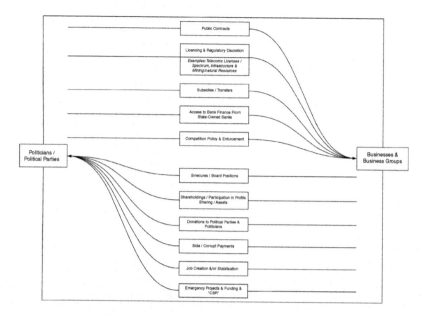

FIGURE 1.2 Channels of interaction between business and politicians

subsidies or transfers. This is often through the channels of state-owned banks with the consequences commonly including the accumulation of large portfolios of non-performing loans and other impaired assets by those banks. As such, state-owned banks with large market shares have over many decades made lending decisions that do not reflect market-based criteria. This has been a pronounced and hugely recurrent feature in the Indian subcontinent, but also in China.

Private businesses have used connection to politicians and political authority also to influence the regulatory context and, sometimes, even the regulatory framework. Telecoms has, notably, been one area where this has been a major feature. For example, in India, the mobile company that has captured the largest market share – Reliance's Jio – has deftly attained that position in part through helpful treatment by the regulators, including the terms on which crucial spectrum and operating licenses have been obtained. In the more tightly controlled economies with a dominant Communist Party, party membership has also proven to be a useful way of facilitating support, whether financial, market access or other. Further, in China there is a common pattern among successful companies. Some have started originally as SOEs, but as the state's share has been diluted over time, former managers and insiders have emerged as the dominant players in the company. The ZTE Corporation is one of the more notable cases in point. Even when a company has been ab initio private, such as Huawei, its founder's close connections to the Communist Party and the People's Liberation Army were critical in securing finance from state banks, contracts with public agencies as well as protection from competitors.

The leading new generation companies in China – such as Baidu, Tencent Holdings and Alibaba – also have very strong links to government which may comprise access to finance but, more often, takes the form of protection from competition, including foreign competition. Reciprocation, not surprisingly, takes the form of compliance with government's preferences and overall objectives. Irrespective of the type of political regime, there is a clear trade-off for connected businesses. When asked by government to take specific actions or finance specific projects, they will always oblige even when it runs counter to their immediate financial interests. Connected businesses are also expected to rise to the occasion at times of national emergencies. It is no accident that across the region, the ravages of COVID-19 have induced many declarations of financial and other commitments by leading businesses to health care or other public agencies. In the Philippines, for example, two of the most powerful business groups – Ayala and First Pacific – made large emergency donations to hospitals and health workers that led the country's president to declare a cessation of hostilities on his part towards these companies.

Aside from listing the channels of interaction, Figure 1.2 also highlights the way in which these interactions tend to have some reciprocal component, whether of a financial or other nature. It further suggests that some sorts of activity are more prone to reliance on connections. In a nutshell, any activity that needs a license or is subject to a high degree of regulatory interference – such as ports, roads and infrastructure projects as well as telecoms and, quite frequently, media – is far more likely to be marked by close connections between politicians and businesses. Similarly, companies in sectors or locations that have large numbers of employees – the scale effect – tend to have a degree of importance that smaller companies lack, and this commonly attracts the attention of politicians. That is because they can be useful in creating jobs, either for themselves, their friends or relations or more generally by boosting employment at propitious times, such as elections. Conversely, the owners of large enterprises can use the fear of layoffs to extract subsidies or other forms of support from government or government-influenced bodies. For example, the Chinese steel industry has been plagued by overcapacity. In cities, such as Wuhan or in the Northeast of the country, excess employment has consequently emerged. Yet, in a sign of sensitivity to unemployment, limited layoffs have been pushed through with steel companies' finances being propped up by public money, including through mergers.

Among the most significant manifestation of the influence of connections has been the way in which particular companies or conglomerates have acquired market power and a resulting attenuation of competition. Further, competition agencies have rarely had the independence or clout to rein in market dominance. And once in that position, it is very striking how adept they have proven at entrenching themselves by leveraging connections to erect or raise barriers to entry. Although competition in external markets has been a salient feature, even export sectors have seen multinational enterprises (MNEs) coexist or partner with powerful, local companies often with the support – explicit or implicit – of government.

I.4   THE CONNECTIONS WORLD IN ASIA:
       PAST AND PRESENT

The idea of connections conjures up images of cronyism and corruption. Both sorts of behaviour – bearing in mind that they tend to be intricately related – have certainly been important features and have been amply documented. For instance, it has been recently argued that the current Chinese system is largely based on crony relationships,[6] many of which have outright corrupt consequences. Consider the case of Liu Zhijun, a former minister in the Chinese government. He was found guilty in 2013 of amassing over $250 million in bribes as well as a huge portfolio of properties and large amounts of foreign currency, not least through the allocation of public contracts. Yet, that level of peculation pales in comparison with the now infamous 1MDB scandal, a government-run strategic development company of that name. In this instance, it is alleged that the then Malaysian Prime Minister – Najib Razak – siphoned off around $700 million into accounts that he controlled, while the main co-conspirator, a Malaysian financier by the name of Jho Low, a man with close ties to China, diverted more than $4.5 billion into opaque financial structures. While the principal players in the scandal were Malaysian, international financial institutions, such as Goldman Sachs, as well as officials from some Arab countries, were also implicated.

Although this scandal seems peculiarly egregious, it is no exaggeration to say that just about every country that we are covering has had in recent times multiple instances of financial scandal resulting from an undesirably close proximity of politicians and business. In the worst instances, it appears that those connections are simply endemic. Of course, these features are by no means particular to Asia. A former president of South Africa – Jacob Zuma – has been accused of multiple illicit and corrupt dealing, including with one family – the Dubai-based Guptas – that facilitated what many have called state capture; the massive subordination of public resources to private interest. Virtually all recent presidents of Peru – as well as many politicians throughout Latin America – have been indicted or convicted for accepting bribes from a major Brazilian construction company – Odebrecht – in return for large public contracts. This sorry

---

[6] Pei (2016).

litany of greed and the – often illegal – subordination of public resources to private appetites cuts across many countries and regions.

Further, copious as these examples may be, it is important to recognise that there is undoubtedly little new in such behaviour. Some of the protagonists clearly follow in the footsteps of politicians who have conflated public and private interests and translated their position into wealth over the centuries. Nor would it be surprising to find that particular companies or individuals and the state have long been closely – and often unhealthily – entwined. After all, most European powers in the medieval period largely allocated land and then trading rights to individuals or families connected to the ruler. The Venetian Empire, based as it was on trade, was organised around its patricians reserving the main opportunities for investment in long-distance trade as well as its supervision and protection. Even the Arsenal – Europe's largest hub of industry in the fifteenth century – had exclusive and prescriptive rights assigned to specific patricians.[7]

In yet more accentuated form, the South Sea Company, set up in 1711 as a joint stock company, was allowed to participate in the management of Britain's national debt and was not only granted trade monopolies but also ensured that an Act was passed which meant that the creation of any other joint stock company would require a royal charter. The bubble associated with its name is emblematic of these moral and economic entanglements. Another British behemoth – the East India Company – came to exemplify the conflation of political power with commercial interest while being beholden only to its shareholders. Following Clive's victory over the Nawab of Bengal and his French allies at the Battle of Plassey in 1757, the spoils flowing to the East India Company were a staggering £232 million and Clive himself pocketed some £22 million.[8] After the East India Company's absorption into the British Empire just over a hundred years later, the colonial system continued to pursue policies that granted explicit preferences for a significant variety of activities, whether of a trading

---

[7] Chambers (1970).   [8] Dalrymple (2019).

or manufacturing nature, to specified companies or trading groups. Part of that preference was founded on race, resulting in indigenous manufacturers and traders being confined to specific niches or struggling to attain any scale.

The examples mentioned previously are mostly drawn from periods prior to the modern heyday of capitalism when mercantilism was the dominant economic model or where political power was organised around imperial dynasties or weakly representative forms of political organisation. These practices became outmoded in the Western economies as industrialisation and its corporate counterparts, along with more democratic political structures, emerged. The resulting changes included rules for ownership and governance and the emergence of the publicly listed corporation and the oversight that this implies. In a nutshell, industrial capitalism in the West fused with institutional formats, notably the limited liability company, that allowed issuance of equity and at the same time provided protection for shareholders. The birth of the modern corporation surely took somewhat different formats depending on each country's legal system, but a common element was the extension of ownership to wider circles than family. Although, such innovations by no means solved tensions between ownership and control rights and the alignment of the interests of managers, owners and shareholders, they created a legal and organisational bedrock that is still largely in place to this day. Yet, what is striking is that many of the patterns of behaviour and organisation that existed in these earlier arrangements have either spilled over or adapted to modern times in Asia. This suggests not only some form of deep resilience and capacity to mutate but also a fairly radical departure from many of the changes that have been introduced by industrial capitalism and its successors in the advanced world. Intriguingly, this different model has developed almost contemporaneously with the emergence of challenges to the Western model of wide share ownership and democracy.

As a consequence of this resilience, in Asia companies – especially those that have major market shares – still tend to be dominated by families or dynasties. Moreover, many are maintained as diversified conglomerates; an organisational arrangement that is a comparative rarity in the advanced economies. These specifics are no accident. They result not just from the relatively recent origin of many of these companies in family ownership but also from the way in which such organisational formats are shaped by, and in turn shape, the highly personalised and connections-based worlds of businesses and politics. Rather than stand-alone, externally owned companies with relatively focussed business or sectoral interests, as predominate in most advanced economies, private, usually family-owned, companies tend to be built around networked organisational formats, such as business groups.

Although the widespread presence of family-based businesses and business group structures can in part be traced to issues of trust and institutional limits on what can be contracted (let alone enforced), it is also very much to do with the nexus between business and politics. As noted earlier, businesses court politicians for privileged access to assets or resources, contracts, priority in public procurement and other preferences, not least limitations on competition. At the same time, politicians promote links to business for a variety of reasons that often include securing income or campaign contributions. Politicians also often demand reciprocity from those businesses that benefit. Sometimes, companies have to pay off debts or other obligations incurred by those to whom they are connected.

This complex skein of interactions and reciprocities makes business groups an attractive organisational format for their owners as they provide suitable vehicles for risk sharing, opacity in accounting and transactions and, crucially, for bargaining with politicians. Their scale and complexity can also act as a deterrent to politicians trying to expropriate or dilute their interests if, or when, they fall out of favour or there is some sort of regime change. And, of course, the very perpetuation of these organisational formats is also a reason for why market failures and institutional weaknesses persist over time.

In sum, the connections world in Asia is testimony to the power of networks and the pervasive, resilient, elastic boundaries between governments and businesses. Some of the links between politicians and businesses are clearly corrupt; others hew a more ambiguous line. Public institutions – such as those concerning law, competition and regulation – have tended to bend to the will of entrenched interests or have even failed to materialise. Yet, viewing these relationships simply through the prisms of corrupt practices and cronyism is too simplifying and obscures the benefits as well as the depth and resilience of the wider connections world. It also fails to capture how connections infiltrate ways of doing things throughout the economy, affecting not just how businesses and politicians interact but also how businesses themselves are set up and function. The implications of this are that the connections world has a very substantial impact on the structure of the economy as well as on its productivity. The economic landscape tends to be dominated by collusive and rent-seeking corporations, individuals and their networks. In a nutshell, connections and the networks on which they are based do not just characterise the present but actively ensure that the future will retain some similar features. As we shall argue, while connections may not have inhibited growth and development in the past – and may, indeed, have sometimes actively helped – that is less likely to be the case in the future. Because Asian capitalism is largely *not* arms-length and impersonal but strongly based on networks, a competitive, popular capitalism has, for the most part, yet to emerge.

1.5 FALLIBILITIES OF THE CONNECTIONS WORLD

Will the connections world and the shapes of the capitalism that it has spawned be sustainable, both economically and politically? Will the power of entrenched networks subvert dynamic processes of innovation and reinvigoration? Will these systems have sufficient adaptability to shocks to retain support from their populations? Moreover, can this broad model of capitalism prove sufficiently dynamic to engage with the sorts of powerful and disruptive challenges that are already present, perhaps most notably those emanating from technological change? These are the questions that this book addresses.

But before going further, it is time to delineate briefly some of the main risks associated with the connections world before providing in the coming chapters a more detailed examination. At the end of the book, we return to these risks, this time with a view to identifying the policies that could address the problems that we have identified.

The connections world, effective and resilient as it has been, nevertheless contains multiple fallibilities, many of which are, in a variety of ways, already visible and some of which are likely to pose substantive risks, possibly of a systemic nature, in the future.

The first fallibility flows directly from the inherent nature of the connections world with its concentrations of influence, power and economic benefits in the hands of relatively small numbers of networked persons, families and, even, dynasties. Important manifestations are the high – and often growing – levels of inequality in both income and wealth, let alone frequent manifestations of corruption. Without a shift to a more open, popular and inclusive form of capitalism, these inequalities will prove debilitating and affect stability.

The second concerns the consequences of the market power and restraints on competition, especially in domestic markets, that flow from the primacy of connections and the resulting weakness of regulatory and other institutions. In most Asian contexts, there is plenty of evidence that connected parties, often based in business groups, work very hard to ensure that they are 'better' connected than others. There is often acute competition among connected entities. Why then does this not necessarily result in undermining the system? The answer is that most of this competition results in the reallocation of rents and privileges rather than their elimination or reduction. As such, the aim of this competition is not really to open up markets or induce the entry of new players, quite the contrary. Barriers are set up against new entrants and disruptors. The principal aim is to redistribute the cake between the incumbents. Furthermore, because success commonly stems from connections rather than education or effort, this provides weak foundations for growth that is based on productivity, creativity and innovation. In some cases, this has actively deterred inward investment or the return of individuals with new ideas and skills from abroad.

The third concerns an important corollary of the connections world; its ability or otherwise to create jobs and achieve an essential – probably the primary – aim of societies; the creation of sufficient and productive employment. It is not the case that it does not create jobs; it does. Many are indeed good jobs, well paid and productive. But the absolute number still remains quite small. The informal economy continues to account for a large – often the largest – share of employment in the Asian economies. Moreover, politicians also often ensure that connected companies, and particularly SOEs, are funnels through which employment, including job creation sensitive to the political cycle, is maintained. The broader implication of this is that much of the necessary employment creation occurs in the informal economy where fragility and low pay are pervasive features. Most troublingly, connected companies are also an important reason for why the formal economy has failed to raise its share of total employment in a major way over time.

The fourth fallibility is closely related. Because padding employment in connected companies and/or SOEs is the dominant mechanism for dealing with adverse economic developments or shocks, the scope for introducing more efficient – and ultimately longer lasting – policies has been far more limited than is desirable. It is striking that as China's average incomes have risen, dealing with individual or households' employment and income risk through creating mechanisms of social insurance has been avoided. Part of the reason for this is that politicians still prefer to rely on mechanisms that they believe are more responsive to their demands and interests. As a consequence, using SOEs and connected companies to cushion labour market risk has remained the dominant approach to the problem. Further, faced with rapid technological change (such as advances in AI) and significant exposure to international markets and value chains, Asian companies have faced pressure to shift more towards capital-intensive production. Absent any effective social insurance, such a shift would imply that citizens in the future will carry significantly higher risk themselves. Moreover, although technological advances can offer the opportunity for new players to challenge incumbents, for this to materialise will require that the space for competition does not get squeezed out by the latter.

The final fallibility concerns political systems and their associated institutions. While, in principle, elections and political turnover are the main ways in which democracies handle change, a significant number of Asian economies are either autocracies or heavily managed democracies. In these instances, the risk of competition among the connected spilling into disruptive turmoil is far greater, if only because of the difficulties in resetting the political equilibrium. Such risks, as a result, are potentially destabilising for both the Chinese and Vietnamese political systems and their associated economic configurations. The lack of adaptability has – both in Asia and elsewhere – been a prelude to disorderly, sometimes chaotic, responses to pressures.

These fallibilities by no means sound the death knell of future growth. Yet, they do highlight that while the connections world has played a central part in Asian success, it also bears the seeds of major challenges in the future. Asian economic performance in the coming decades will depend on how countries respond to these challenges and their capacity for adaptation. In the final chapter, we ask how policies could begin to address some of the fallibilities that we have identified. Whilst clearly not straightforward – due to the entrenched behaviour and advantages of the connected – we suggest specific changes that could be put in place to counter the market power accumulated by business groups along with the allied consequences of the limited creation of productive jobs, let alone the exacerbated levels of inequality that have emerged. Such changes could also radically influence the governance of companies and in so doing achieve far higher levels of transparency and accountability. Moreover, it could create space for the introduction of more modern systems of risk management, both for workers and companies. This would limit the need to rely on connections and reciprocity-based deals that characterises the present-day connections world.

1.6   A GUIDE TO THE COMING CHAPTERS

In the next chapter, we provide a building block for the analysis which follows. We start by setting the context, looking at Asia's economic ascent and the factors behind it. That ascent has shifted its share of the world economy from 9 per cent to nearly 40 per cent over the last half-century and seen income levels and living standards rise dramatically. We use data on the economy, political systems and institutions from a variety of sources to chart the progress of Asian countries since the 1970s. What becomes clear is that despite this ascent, the gap between incomes in most of Asia and those in the rich world are still huge so that there is considerable convergence yet to be achieved. Moreover, productivity levels are but a fraction on average of those in the economies of Europe and North America. For these reasons, the ascent – though steep – has realistically only taken Asia to a staging post for any future drive to the peak. Further, growth has been mainly driven by population increase and investment, financed by high savings and, particularly in East Asia, export-led integration into global value chains (GVCs). Yet, some of these factors are now less favourable: the demographic dividend is being exhausted and globalisation is being challenged, a process being accelerated by COVID-19. In addition, whilst poverty has declined, economic insecurity remains ubiquitous and inequality has risen steeply. Although the quality of formal institutions has often improved, non-market supporting informal institutions remain significant almost everywhere and many countries remain wedded to autocracy or forms of 'managed' democracy. In short, for Asia to go from the foothills to the summit will require not only a shift away from the extensive growth model but also a profound re-evaluation of the organisational, institutional and other arrangements currently in place.

Chapter 3 turns explicitly to the way in which Asia's connections world is configured, highlighting the extraordinarily pervasive nature of ties between businesses and politics and the networks on which they are based. These networks derive advantage from their mutual – often reciprocal – relationships. Most of these relationships are strongly transactional but they also affect how individuals and companies actually organise themselves to achieve some of these goals. For example, the institutional framework for private companies is often designed with a central – even primary – purpose of leveraging resources and assets, as well as gaining advantage, whether in relation to the regulator or with regard to other actual or potential competitors.

The nature of these connections and their associated networks is initially described using information on politicians, political parties and various types of businesses for each of the countries on which we are concentrating. We rely on a novel dataset that puts together comprehensive information on politically exposed persons and institutions in all of Asia. This information allows us to map the various networks at the level of each country. These maps highlight significant differences between countries, mainly resulting from the variation in political systems and related institutions. Although these maps provide a useful starting point, to understand how connections actually play out requires using detailed cases and examples from across Asia. Once done, we find that that whatever the local variation, these webs of connections bind together with common purpose. Moreover, leveraging connections for mutual benefit has often delivered large and very enduring benefits that have proven resistant to changes of government or even political regime. Indeed, in today's Asian democracies, many large, sometimes dominant, businesses have been built on the connections established in earlier autocratic eras. Perhaps most importantly, these webs of connections have created a system of behaviour that has become increasingly entrenched whether in the richer economies, such as South Korea, or in some of the poorer ones, such as Pakistan. Such behaviours also cut across political systems.

The connections world not only influences in important ways how businesses organise themselves but also how they function.

Chapter 4 examines a central, indeed a defining, feature of the Asian economies: family-based and founder-manager business groups. These comprise networks of firms bound together through formal and informal ownership links, with a family or dynasty usually at their heart. Some are truly massive by global standards; most are highly diversified, and they are often the dominant players in their home country across a wide variety of industries.

What we find is that business groups are a uniquely well-suited format for doing business in the connections world. Opaque cross-holdings and pyramids of stocks ensure that families can exert effective control, even if their actual shareholdings are relatively small. These arrangements also open up endless opportunities for playing reciprocity games with politicians, civil servants and members of other oligarchic dynasties. Although there are examples of efficient and well-run business groups, most do not conform to this characterisation. Furthermore, while it has often been argued that business groups are a response to institutional and market weaknesses – for example in relation to securing finance – they have not faded with growth and the improvement in institutions. Rather, business groups have become more entrenched in Asia over time.

Just what has happened is shown by our measure of economic prevalence. This indicates the impact of the largest business groups on an economy in terms of their market power. In many Asian economies, half a dozen or fewer business groups generate revenues which constitute the majority of the country's economic activity. Such concentrated ownership has also had an impact on extreme wealth. The growth in the number of billionaires has been quite staggering – from only 47 in 2000 to 719 twenty years later! This entrenchment of economic power and wealth underpins the operation, and reflects the consequences, of the connections world.

With business groups and connections playing such a major role throughout Asia, what are the implications for making the transition to growth based on technological advance and innovation? In Chapter 5, we explore how these economies sit globally in terms of innovation. What transpires is that most Asian economies are not very innovative by international standards. Most perform in line with what could be expected for their level of development. Asian economies mostly obtain their technological innovation from abroad through foreign direct investment (FDI) or by domestic firms obtaining these technologies in nefarious ways from licenses to imitation. However, despite very rapid growth – and with the exception of China – the attraction of FDI by the Asian economies has actually been distinctly lacklustre. Part of this is the consequence of the connections world. Politicians and business groups have been mutually supportive in erecting barriers to entry by new firms, be they from abroad through FDI, or from domestic entrepreneurs trying to disrupt domestic incumbents. As a result, most innovation has been within business groups or by new firms entering new sectors where existing business groups were absent or had not managed to erect unscalable entry barriers.

However, even if most Asian economies will not be able to base their growth on their own innovations in the near future, there are three countries which have developed a base for innovation: China, India and South Korea. In each, long-standing efforts to construct an environment favourable to innovation, including policies for education, science and technology, as well as encouraging returning migrants with knowledge, are beginning to reap dividends. Each has adopted a rather different model, generally centred in business groups. In India, this has mostly happened where the dead hand of state regulation and licensing has been weakest. In South Korea, innovation-driven growth is based on a compact that knits together the massive, incumbent business groups and the state. In China – with its aspirations for being a global technology leader – the onus has shifted to notionally private companies spearheading the effort but with substantial support – financial and otherwise, along with increasingly vigilant – not to say, intrusive – oversight from the autocratic state. Yet, even despite these achievements, the power and influence of the connections world in these three countries also remains a serious brake on their ability to innovate in the future.

Although much of Asia's success in recent decades has been focused on growth, governments actually worry just as much about the amount of employment that has been created. Presently, just to keep the share of employment stable, Asia needs to generate over a million jobs a month. In Chapter 6 we examine how well these goals have been achieved and the forms of job creation that have resulted. We show that, even when employment targets in aggregate have been achieved, the ways in which that has occurred mostly fly in the face of governments' declared objectives. Specifically, most employment remains in the unorganised or informal parts of the economy. Those jobs are generally fragile, low wage and low productivity. Although public sector employment is often quite high, the reduced enthusiasm for SOEs throughout Asia has meant that 'good jobs' are now mostly to be found in the private sector. For sure, many of the business groups that figure in the connections world also create productive and relatively well-remunerated jobs. As do other large companies, including foreign-owned ones. The problem, however, is that the organised or formal private sector is of limited size and lacks the ability, or even willingness, to create substantially increased numbers of jobs. In the connections world, business groups and other established companies may compete with each other, but the scale of entry and exit, as well as rivalry, is held in check. This has led to a pronounced polarity between working conditions, compensation and productivity in the numerically dominant informal firms compared to the relatively small number of larger formal ones. Boosting formality and, with it, productivity – a clarion call of almost all Asian governments for decades – has largely failed to materialise.

The entrenchment of the connections world has also helped ensure that little or no progress has been made in bringing in more effective responses to employment risk. In this world, neither government nor companies have a strong interest in promoting arms-length methods of dealing with such risk. They prefer, rather, to rely on discretion. Jobs can be created, and their destruction tempered, as a result of interactions or even haggling between politicians and employers. The bulk of workers, namely those who function in the informal economy, are de facto excluded. And the modernisation of welfare systems – now feasible given the income levels of China, Malaysia and others – remains stalled.

Our final chapter places our argument in a longer perspective. The aim is not to make prognostications about the future of each of these countries. But it is to say something about how the characteristics of today will affect the paths of Asian growth and development tomorrow and, in particular, to draw conclusions about what will be the impact of the connections world on those prospects. Despite many similarities, the connections worlds still have many local features, dynamics and, ultimately, prospects. Even so, set against the tasks of climbing the steep income mountain towards convergence with the rich world and doing so in ways that do not necessarily confer disproportionate rewards on limited groups of people and institutions, the connections world imposes some quite common constraints.

One is the ability of powerful businesses and families to entrench themselves by virtue of their connections to government and/or politicians. This is as true in China as it is in India. The mutual benefits of the current system for politicians and business owners mean that neither side has much, if any, incentive to move away from the current arrangements. Hence, limiting competition and suppressing the dynamic processes of entry and exit, particularly in the formal or organised part of the economies, will be hard to shift. We propose a series of measures and policies aimed at improving transparency and governance more generally, whether in firms, capital markets or in relation to government and politicians. Because marginal changes pursued through existing competition or regulatory authorities are most likely not to be credible or effective, we propose a set of more radical measures to disrupt and refashion the connections world that, ideally, should be taken simultaneously. Among them are the use of prohibitions on cross-holdings and other business group practices, as well as the application of tax reforms, including the introduction of inheritance taxes, that can begin to shift the balance of advantage away from the business group format. In addition, we propose a series of changes aimed at boosting competition and ultimately making competition policy more effective. At the same time, improving political transparency and oversight, so as to limit the incentives for politicians to perpetuate the connections world will be essential.

Finally, we focus on the main pressure points to which the connections world is, and will be, subject. As signalled earlier, these are the ability, or otherwise, to generate innovation through reliance on new entrants and an effective entrepreneurial ecosystem. With a few notable exceptions, this has so far proven elusive, not least because of the entrenched market power of business groups and their cossetting by government.

Then there is the constant pressure to create sufficient jobs. Not only has the connections world been actually quite poor at creating productive, 'good' jobs but technological change is beginning to bite and will affect both the level and type of employment. In some countries – notably China – those changes have paradoxically been encouraged with a view to gaining strategic advantage in AI. But whatever the context, the reality is that if stuck with their current arrangements, the Asian economies will struggle to satisfy the demand for jobs. That will likely mean further increases in the size of the informal economy. This will doubtless be accompanied by continued reliance on subsidies to preserve some part of employment in the formal economy, rather than the creation of better systems for managing employment risk.

An additional pressure point comes from ballooning inequality of income and wealth. High inequality tends to be associated with economic underperformance let alone susceptibility to political turmoil. This is especially problematic with autocracies. While progressive taxes and greater coverage can help mollify inequality, that has to be led by effective targeting of the sources of that inequality, not least the features of the connections world that this book will have described.

Before we get going, let us emphasise that Asia's ascent – and the wider ripples that it has induced across a broad swathe of emerging economies – has been testimony to a remarkable marshalling of resources and, in some instances, highly effective public policy. Many households have been pulled out of poverty and income gains have been substantial, if unequal. These are huge achievements. But it has also revealed fallibilities, not least the accretion of market power and an unhealthily close relationship between business and politicians. What we have termed the Asian connections world has proven very effective in limiting possible encroachments on the privileges that it has secured. However, the organisational forms of this model of capitalism – notably business groups and state-owned firms – exhibit features that betray not only their purpose but also their weaknesses.

2   To the Foothills of Everest
Asia's Resurgence

2.1   INTRODUCTION

Most discussion of Asia's place in the world economy alights almost effortlessly on the continent's extraordinary recent growth and the associated decline in poverty that has also ensued.[1] The story is not only a stock-in-trade of countless analysts, but also plays a prominent part in the narrative of Asia's politicians. Political leaders vie to exceed each other in the annual growth horse race, even if China has in most years been the odds-on favourite. Many leaders have spent heavily on engaging advisers, and public relations companies, as well as enlisting international organisations, to promote their various 'visions' for their countries along with the growth targets that generally take centre-stage.

Encapsulating Asia's recent past in its record of growth requires no adulteration of facts or violation of realities. The Asian economies have also emerged onto the broader global stage, whether in terms of their share in global trade or in wider geopolitical terms. Among these many expanding magnitudes, Asia now accounts for over a third of the world's merchandise exports with its overall trade flows having increased by more than one and a half times just in the last decade.[2] Within this broad regional aggregate, China, Vietnam and India have been the most dynamic in terms of the growth in trade, testimony in part to the role of regional value chains. But it has not necessarily been a simple climb; there has also been some recalibration. China's focus on exports has shifted. Between 2006 and 2018 exports as a share of

---

[1] There are numerous studies that chart Asia's economic progress in past decades, see for example Schipke (2015), World Bank (2018a, c), Mason and Shetty (2019).

[2] WTO Statistical Review (2019).

the country's GDP pretty much halved.[3] That has not stopped China from being a powerful force in international trade, but the emphasis has been increasingly on higher value-added products as well as on the domestic market – a feature that we discuss in more detail later in the chapter.

With this enhanced trade presence has also come a higher profile in investment, both in Asia and outside. The largest spending on infrastructure projects across many continents is now being financed within the framework of China's Belt and Road Initiative (BRI). MNEs – particularly from India and China – have acquired assets through the region and expanded their presence in markets.[4] FDI into the region has grown rapidly. In 2018, there were seven Asian countries in the global top twenty economies for receipt of FDI, including China, Hong Kong, Singapore and India that were in the top ten. These huge inflows have played a major role in transferring technology and knowhow, improving working practices and, in some instances, have led to significant improvements in productivity. For example, substantial growth in Thailand's labour productivity has been largely fuelled by inward FDI, involvement in global supply chains and export markets. As a result, the country's mix of industries has also moved closer to that existing in the advanced economies. At the same time, Asian firms have become leading multinationals in their own right, think, for example, of household names like Samsung, Tata, Huawei and Alibaba. China is now home to more than one hundred of the world's largest firms, compared with 134 in the United States.[5] These large firms increasingly operate in many jurisdictions. Indeed, among the top twenty countries in terms of FDI outflows in 2018, six came from Asia including again near the top, China and South Korea. Beyond simply the economic dimensions,

---

[3] Falling from 35 per cent to 18 per cent.

[4] In 2018, 42 per cent of all FDI originated from developing countries, as against 25.6 per cent in 2010 and only 7.6 per cent in 2000 (UNCTAD, 2019, 2020).

[5] Fortune (2016). China now has 103 of the world's largest firms as against 134 in the United States and 142 in the EU. South Korea has 15 and India 7.

both India and China have also actively tried to project soft power[6] as they search for greater influence in the region. Whilst the smaller economies of Asia have played less prominent roles, many have emerged as significant trading and investment partners of their giant neighbours.

This chapter is a selective account of how and why the share of Asia in the world economy has more than quadrupled in the past half-century. Selective because there are already many standard accounts of the rise of Asia – especially China[7] – but also because we highlight issues less emphasised by others but relevant to our narrative about the consequences of the connections world for long-run economic growth. In that sense, we follow Myrdal (1972, ix) in arguing that 'economic problems cannot be studied in isolation but only in their demographic, social and political setting'. Our main focus is characterising Asian growth, which has occurred rapidly throughout the region regardless of political system, institutional arrangements or policy cocktails. We also illustrate how far the Asian economies have come and how far they have left to go to attain the living standards of Europe or North America. In particular, we concentrate on the features that are both a cause and a consequence of the connections world. These include political arrangements and economic institutions, but also inequality. In so doing, we set the scene for the chapters that follow.

2.2 GROWTH IN ASIA

2.2.1 *Growth and Political Systems*

The main big surprise is that Asia's growth has not just spilled across borders but mostly across political systems as well. For sure, as will become clear, there are some within-region differences, but it is safe to say that we are talking about a continent-wide phenomenon that

---

[6] And sometimes hard power as the disputes over South China Sea islands indicate.
[7] See, for example, Maddison (2007), World Bank (2018a, c) and Mason and Shetty (2019).

has occurred against a backdrop of very different political systems, along with substantial variation in the institutional arrangements that accompany those political systems. Using a classification of political systems over time, we can see that Asia embraces a wide diversity, ranging from the competitive democracy of India to the largely autocratic regimes of China and Vietnam, but also Thailand. There have also been significant changes to countries' political systems over time. South Korea now ranks as an open democracy with regular turnover of governments and separation of powers. Yet, that country's economic ascent in the 1970s started with military regimes and a harsh autocracy. Similarly, the extent of autocracy has varied within periods even in countries, such as China and Vietnam, where it is still the dominant form today.

A simple but informative way of illustrating our point is contained in Figure 2.1 which provides a scatter of average growth by a country between 2000 and 2018 against the median score for its

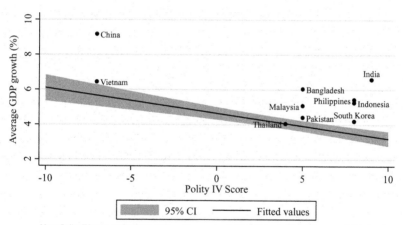

Note: Polity IV score ranges from -10 (hereditary monarchy) to +10 (consolidated democracy). It is based on regulation, competitiveness, openness of executive recruitment, executive constraints, regulation and competitiveness of participation. Polity IV Score is the median of the revised combined Polity IV Score in time period 2000-2018. Average GDP growth (%) is the mean of GDP growth (annual %) in time period 2000-2018.
Data source: Real GDP growth (annual %) in 2010 US$- World Development Indicators (WDI), Polity IV-Center for Systemic Peace (CSP)
Number of countries: 162

FIGURE 2.1 Relationship between GDP growth and political system, 2000–2018

political system for the corresponding period.[8] The possible scores range from -10 for a strong autocracy to $+10$ for a robust democracy. It will be seen that China and Vietnam score fairly low using this measure of political system, while the remaining countries are all on the positive side of the scale (democracies) ranging from weaker ones, such as Thailand and Pakistan, to stronger ones, such as South Korea, Indonesia and India. To place this in context, a regression line for 162 economies worldwide, with a 95% confidence interval (CI), is also included. This allows us to consider each Asian economy relative to the rate of growth predicted for a country with that political system using data from most of the countries in the world. Thus, deviation from the regression line indicates the difference between that country's growth rate and the global norm for countries with the same political score. The 95% CI enhances our understanding of the deviation from the global norm by providing the range of values of growth that we can be 95% confident will contain the true mean of the underlying global set of countries. As such, a country which is outside (above) the 95% CI is growing faster than we would expect given its political system. Looking at the countries of Asia, it is evident that growth rates between 2000 and 2018, for all except Thailand, are significantly above the regression line. Unsurprisingly, perhaps, China and India lead the pack, but many countries average above 5 per cent growth. At the same time, we find in Asia that rapid growth is not the prerogative of any particular political system. Indeed, with the exception of China, which is both the most autocratic and the fastest growing, there is no discernible relationship between political system and economic growth in Asia.

This contrasts with the rest of the world, represented by the regression line. Globally, the relationship between autocracy and economic growth is mildly negative, although, of course, this association is not necessarily causal. Poorer countries are perhaps more likely to have undemocratic regimes but also on average to grow faster

---

[8] Marshall et al. (2019).

because of their low initial levels of income and hence space for faster growth and convergence. Before looking into how this rapid growth has been generated and as to whether it is likely to persist, it will be helpful to start with a rapid description of Asia's economic history over the past half-century.

2.2.2   Growth – and Lots of It

In 1500, India and China had – as best can be measured – broadly comparable levels of income per capita as Spain and were not far below levels obtaining through most of Europe[9] Yet, by 1800 most European nations were pulling away and had incomes roughly double those of the large Asian economies. Colonialism along with the Industrial Revolution ensured that the gap only widened. By 1950 average incomes in Europe were between four and fourteen times those in China or India.

There is much that is striking in this discrepancy, none more so than the fact of Asian incomes being becalmed at a low and stable level for nearly five centuries. Yet the second half of the twentieth century has seen this vast continent shaken out of its income torpor. Yesterday's developing countries have swiftly trans-mogrified into today's emerging markets and, in some cases, such as South Korea and Singapore, into advanced economies with high levels of average income. This in turn has dramatically boosted Asia's share in the world economy. In 1970, Asia (excluding Japan) accounted for around 9 per cent of the world economy. At the turn of the twenty-first century this had climbed to 18 per cent and today exceeds 40 per cent.[10]

This dramatic change in pace has – particularly since the 1980s – been visible through most of Asia. There have however been significant differences within the region. In South Asia, the ascent has been slower. Although both India and Bangladesh have seen their average incomes (expressed in constant PPP) grow by over 4 per cent

---

[9] Baten (2016).   [10] Maddison (2010).

per annum over last thirty years, in Pakistan growth was just 1.6 per cent. Undoubtedly, the most striking case of a persistent and powerful acceleration in growth has been China. Over the last thirty years, average incomes in China have grown by nearly 9 per cent each year. China's economy is consequently over thirty times larger than it was in 1990 and seventy-five times larger than in 1980, around when some of the serious changes in policy towards a market economy started to be put in place. By 2020 China stands as the world's second largest economy: a feature often trumpeted by the country's rulers. This scale provides China, and to a lesser extent India, with the financial clout to play an increasingly significant role in advancing technology, from AI to space exploration.

Asia's rapid growth has also led to fundamental changes in the global balance of economic power. In 1970, the United States was the dominant world economy producing around 46 per cent of global GDP, and all the other countries in the global top ten were European, except for Japan and Mexico. No Asian economy had a GDP representing more than 1 per cent of world GDP. But by 2020, the US and European shares fell below 50 per cent, while China and India had joined the world's largest ten economies with shares of 16 per cent and >3 per cent, respectively. South Korea and Indonesia had also entered the world's top twenty economies.

2.2.3 Convergence or Not?

Asian economies have also for the most part been growing faster over the long haul than would be implied by their level of development. Economists normally predict a negative relationship between the rate of growth of an economy and its GDP per capita, something attributable to convergence. What this means is that poorer countries could be expected initially to grow faster than their more developed counterparts, with that growth rate gradually decelerating as they become richer. This might explain why, for example, in the past few decades, Chinese growth rates have gradually come down from in excess of 10 per cent to more like 6 per cent.

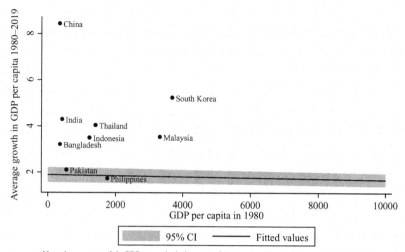

FIGURE 2.2 Average growth in GDP per capita (1980–2019) versus initial GDP per capita (1980)

Figure 2.2, which follows the format of Figure 2.1, explores the matter of convergence. It shows average growth rates in Asia over a long period against GDP per capita in the initial year and again plots a 95% CI against the regression line. Thus, the figure reveals both the growth rates in each of our Asian economies between 1980 and 2019, and how this is affected by their initial level of development (GDP per capita) We can use this along with Figure 2.1, which shows more recent growth rates between 2000 and 2018 to compare recent and more long-run growth rates. From this we can see that average growth rates in countries such as China and South Korea have been higher over the longer period than the shorter and more recent period. In contrast, Bangladesh, Philippines, India and Indonesia all have experienced higher average growth rates between 2000 and 2018 (Figure 2.1) than since 1980 (Figure 2.2), suggesting that rapid growth started earlier in China or South Korea. Figure 2.2 also contains a regression line of growth against GDP per capita in 1980, again estimated using

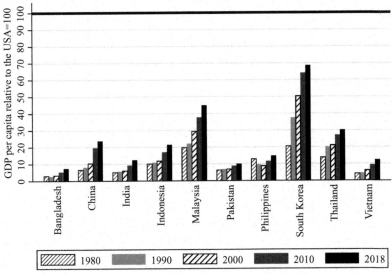

Data source: Real GDP per capita in 2011 US$- Maddison Project Database (2020)

FIGURE 2.3 GDP per capita relative to the United States, 1980–2018

all the countries in the world for which comparable data are available, a total of 138 countries. The regression line is gently downward sloping, consistent with the idea of convergence as rates of growth decline as GDP per capita increases. Moreover, Figure 2.2 makes clear that almost every country in Asia has displayed long-run growth that is faster than might be expected for their level of development. The only exceptions are Pakistan and Philippines and even they are within the 95% CI.

Asia's experience – unlike that, say, of Africa in recent times – shows not only that convergence has been occurring but that it can be reasonably speedy, although this is by no means uniformly true. Figure 2.3 gives the evolution of GDP per capita by decade since 1980, while also showing how each country's income stands relative to the United States for each decade. In South Korea's case it has taken nearly fifty years (1970–2020) for its income per capita to reach roughly the level of Spain and two-thirds that of the United States. Over the same period, Malaysia managed to climb from 16 per cent to

45 per cent of the US level while Indonesia and Thailand went from around 10 per cent to 19 per cent and 30 per cent, respectively. By contrast, the Philippines has been stuck at around 14 per cent of the US level over the last half-century. In the case of the two large economies – India and China – income per capita went from just under 5 per cent of the US level each to around 11 per cent and 28 per cent, respectively, in fifty years.

Even so, the convergence process still has a long way to run, incomes per capita in Asia still remain low. Presently, only Malaysia and South Korea have surpassed 50 per cent of the level in high-income countries or that of the United States. Chinese incomes are barely 36 per cent of the average for the rich world and less than 28 per cent of the United States, while Vietnamese and Filipino incomes are no more than 16–18 per cent and 13–14 per cent, respectively. In the subcontinent, both Bangladesh and Pakistan have average income levels under 8 per cent that of the United States, with India around 11 per cent.

In short, GDP per capita has more than doubled in this period in every country except Pakistan, Philippines and Thailand. As a consequence, most Asian countries started at or below 10 per cent of the US level in 1980 but had attained levels of more than 20 per cent by 2018: growing on average at least double the US rate. Whilst enormously impressive, this still means they have a long way to go – even now only South Korea currently exceeds half the US level of GDP per capita, while Bangladesh, Pakistan and Vietnam are less than 10 per cent.

Another way of looking at the matter is to see whether there has been any convergence in labour productivity. One reason for focusing on productivity is that it will always be the main force behind a narrowing of differences in incomes. Here, the most recent evidence suggests that there has been some convergence since the early 2000s but the rate of convergence to advanced economy levels in both East and Southeast Asia and South Asia has been quite gradual, at between 1.5 and 2.5 per cent, respectively. In short, the gap remains very large and the rate at which it is being closed implies no early convergence.

Whether focusing on measures of per capita incomes or productivity, Asian countries have certainly kick-started their economies after decades or even centuries of torpor and reduced the gap with the rich world. However, a very great deal of catch-up remains to be done. Even if China were to sustain the very high – in the region of 9 per cent – growth rates per year that it achieved between the early 1990s and 2013, it would still take that country at least another thirty odd years to draw close to US income levels per capita. Of course, that almost certainly is highly optimistic, not least because the process of convergence itself will lead to a deceleration of Chinese growth rates. For India, even at current high growth rates, convergence will not happen for at least seventy-five years!

Put in wider context, it is worth emphasising that while episodes of very rapid growth (>6 per cent per annum) have occurred historically, for example in the Soviet Union in the 1930s, such episodes have been mostly quite short-lived. China, until recently, has been the glaring exception, although there are serious doubts about the reliability of its growth numbers. Previous periods of very fast growth have almost always ended in sharp decelerations and reversion to the world mean growth rates.[11] The case of Japan may be instructive. It grew on average at around 8 per cent after the end of the Second World War until the 1970s. By 1973 GDP per capita was 69 per cent that of the United States. Convergence then continued at a slower rate until 1991 at which point Japan's GDP per capita was 85 per cent that of the United States. Subsequently, growth slowed drastically putting convergence into reverse. In other words, not only do current growth rates have very weak predictive power for future growth but there are good reasons for doubting that convergence will be something almost automatic.

Indeed, a variety of arguments have been put forward to suggest that further convergence is likely to prove harder in the future, especially in Asia.[12] One is that the fall in manufacturing's share of output

---

[11] Pritchett and Summers (2013).   [12] See Kohli et al. (2011).

in many emerging economies heralds less scope for productivity and incomes to grow. Moreover, this may be accentuated by changes in technology that introduce greater automation while also reducing the attraction of emerging markets as places for manufacturing. We will return to these issues in Chapter 7. But it is not just the shift away from manufacturing, along with technological change, that challenges the convergence narrative. Institutional and governance factors can also play a major role in holding back the process. Indeed, such restraining factors have even been wrapped up in a term – the middle-income trap.[13] In that view, growth will be hard to sustain due to the absence, or partial presence, of needed and complementary requirements, such as education and skills, as well as secure property rights, and other constraints.

These are all valid caveats. Yet what this sort of argument has ignored so far is that the underpinnings of modern Asian capitalism – notably its reliance on what we term the connections world – are already a challenge to the region's longer-term growth prospects and the sustainability of its broader model. In fact, some of the shortcomings that are wrapped up in the idea of a middle-income trap are themselves the direct products of how the connections world functions. But, before we go in greater depth into why and how that is the case, we need to spend a bit of time understanding some of the main ingredients of the model that has enabled the extraordinary transformation in Asia's growth.

2.3   ASIA'S GROWTH MODEL SO FAR

2.3.1   Industrialisation and Export-led Growth

A major driver behind Asia's rapid growth has been the transfer of relatively unskilled workers from the low-productivity, agricultural sector to an urban, industrial and service sector. The rapid industrialisation of Asia is demonstrated in Figure 2.4 which shows the share

---

[13] Eichengreen et al. (2013); Lewin et al. (2016).

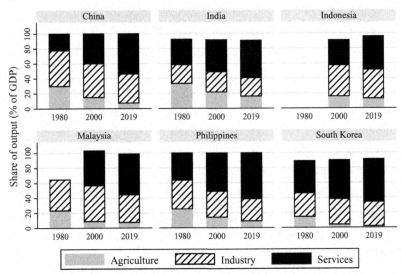

Data source: Agriculture, forestry and fishing, value added (% of GDP), Industry (including construction), value added (% of GDP), Services, value added (% of GDP)- World Development Indicators (WDI)

FIGURE 2.4 Share of output by sector 1980–2019, selected Asian countries

of output in the three main sectors of the economy – agriculture, industry and services – in 1980, 2000 and 2019. This identifies how the industrial structure of each country has evolved over the past forty years. The proportion of output accounted for by agriculture has dropped precipitously everywhere, reaching low levels in China and Malaysia and almost developed economy levels in South Korea. At the same time, industrialisation has raised the shares of services and industry, with the former exceeding 50 per cent everywhere, except for Indonesia. But here is the rub. Figure 2.4 shows that de-industrialisation has already begun to set in, particularly in China and Malaysia, but also more widely. Further, as we discuss in Chapter 6, although agriculture's share of output has fallen substantially everywhere, significant shares of employment are still locked up in low-productivity agriculture in most countries.

World Development Indicators (WDI) calculate total value added of GDP for a country as follows. It combines agriculture, industry and services, excluding financial intermediary services, indirectly measured (FISIM) and for those countries that report value added at basic prices, net indirect taxes are reported separately. As a result, value-added shares of output for agriculture, industry and services may not always add up to a hundred per cent.

Asian growth is also distinguished by being highly export-led, primarily of manufactured products. Initially in Japan, then South Korea and subsequently China and Vietnam, the model of development was therefore strongly externally focused, giving weight to exporting into international markets[14] and integrating into the GVCs of large companies or sectors. For example, the OECD has put together a GVC Participation Index which indicates the extent to which a country is involved in a vertically fragmented production process. They report that South Korea, Philippines and Malaysia have levels of participation in GVCs – in excess of 65 per cent, higher than almost all (other) OECD economies, except Luxembourg. Other Asian economies, including Thailand and Vietnam, are above 50 per cent GVC participation, around the median for all OECD economies. The larger Asian economies – China, Indonesia and India – have somewhat lower values of the Index, at between 40 and 45 per cent.[15] For these values to rise or fall in the future will likely depend on whether a greater focus on innovation will affect the criteria that firms use in choosing locations for their value chains, not least the ability to exploit new technologies rather than reliance on low labour costs.[16]

2.3.2   The Labour Force

Another key element in Asian growth has been a plentiful supply of labour, generated by the favourable demographics that previously existed and to some extent still exist, in South Asia. To illustrate this, Figure 2.5(a)–(l) show population age distributions for four Asian countries – China, India, Indonesia and the Philippines – between 1990 and

---

[14] Amsden (2009).   [15] De Backer and Mirodout (2013); Baldwin (2016).
[16] World Bank (2017).

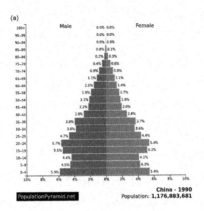

FIGURE 2.5 (a–l) Population pyramids 1990–2019, selected Asian countries

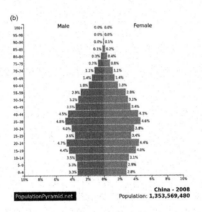

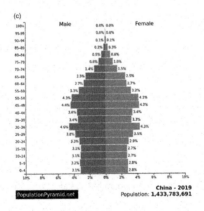

FIGURE 2.5 (cont.)

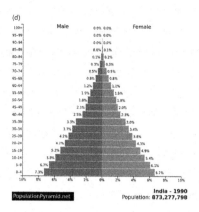

India - 1990
Population: 873,277,798

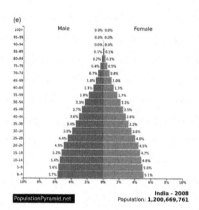

India - 2008
Population: 1,200,669,761

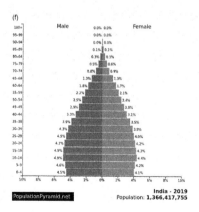

India - 2019
Population: 1,366,417,755

FIGURE 2.5 (cont.)

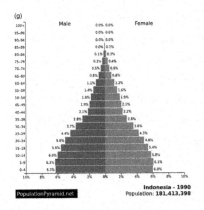

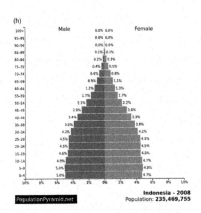

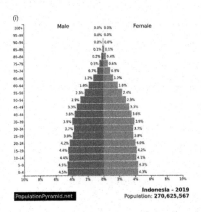

FIGURE 2.5 (cont.)

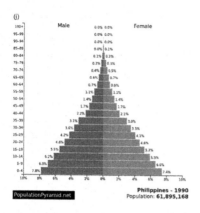

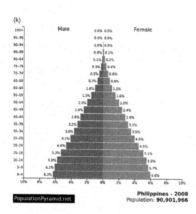

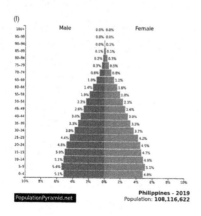

FIGURE 2.5 (cont.)

2019. These show the male and female proportions of the population in age categories from 1 to 100 years old. In countries which are fast growing, young people are a higher proportion of the population, and the figure will therefore look like a pyramid. In 1990, the shape was indeed pyramidical everywhere, except China where the one child population policy was already influencing its demographics. With these favourable demographics of rapid population growth, the labour force expands which facilitates faster economic growth. However, in recent years, the shapes have become increasingly rectangular, consistent with a stable or declining population and an increase in the dependency rate or the proportion of the population who are not working. These distributions are consequently beginning to display the same shape as observed in advanced economies, where population growth has slowed or stopped and countries' average age has been rising. One result is that these demographic trends are starting to put greater pressure on Asia's health and welfare systems. This is happening at much lower levels of GDP per capita than in Europe or North America.

2.3.3   High Savings Rates

Rapid industrialisation in Asia has relied, especially in the most successful countries, on high domestic savings rates – around 45 per cent in China, 35 per cent in South Korea and 18 per cent in India. These domestic resources have been bolstered in many countries by large capital inflows from abroad, including, as we already noted, through FDI. The high savings rates are also a consequence of demographics. In addition to saving for housing, households have accumulated financial resources for when they grow older due to the very low levels of public provision of healthcare and pensions. However, the process of ageing, indicated in Figure 2.5(a)–(l), suggests that these levels may drop substantially in the future. In China especially, as we discuss subsequently, high savings have also been a plank of government policy for financing growth and infrastructure spending. As in the former Soviet Union, suppression of consumption has therefore been a long-term policy tool.[17]

2.3.4 Accounting for Growth

In trying to pull these various ingredients together to identify the dominant drivers behind Asia's improved performance, economists often resort to growth accounting. This is a method of analysis that tries to separate out the different contributions of capital, labour and total factor productivity (TFP) to growth in national income or productivity. TFP is actually a residual – that part not explained by capital or labour – and captures a variety of facets, not just technical progress, but also public policy and other aspects that might affect the efficiency with which capital and labour is deployed.

In East Asia's acceleration in the 1970–1980s, innovation and rising TFP were found to have played only a modest role, with the main drivers being the accumulation of capital and labour along with some improvements in the quality of the labour force.[18] The contribution of TFP, while by no means absent, and also varying across countries, was mostly a much less significant contributor to growth.

The evidence for what has been driving growth after the 1980s points to a greater role for TFP. One study of India and China found that TFP accounted for at least half of the substantial increase in output per worker in both countries between 1978 and 2003–4.[19] When breaking this down by broad sector, the contribution of TFP to industry's output per worker was substantially larger than that of either capital or labour in China but not in India, although TFP's contribution did increase significantly in the latter after 1993. Services contributed more to India's growth with an increase in TFP being the main driver.

Since the early 2000s, productivity growth in East and Southeast Asia has run at around 9 per cent between 2003 and 2008, subsequently falling back between 2013 and 2018 to 6 per cent. In South Asia it went from 6.4 per cent to 5.3 per cent over the same

[17] Ofer (1987).   [18] Young (1995).   [19] Bosworth and Collins (2008).

period. Although decelerating, these rates of increase have been significantly higher than in other regions of the world. Some countries have also done better than others. For example, in the most recent period, Indian productivity has grown by 0.5–2 percentage points more than either Bangladesh or Pakistan. Chinese productivity growth, followed by Vietnam's, has also been the strongest in East Asia.

But even with this strong performance, it must be emphasised that both Chinese and East and Southeast Asian productivity still averages only 12 per cent of advanced economy levels, whilst for Pakistan, Bangladesh and India in South Asia, it is only between 3.5 and 5.5 per cent. To put this in context, from the current situation and with current rates of productivity growth being sustained, the World Bank still estimates that it will take most of East and Southeast Asia's economies up to forty years just to halve the productivity gap with the rich world.[20]

Placed in broader perspective, a major source – possibly as much as 30 per cent – of productivity growth since 2013 in South Asia – has been from the continuing shift out of agriculture into more productive sectors. There is still considerable scope for more of this as India, for example, still has around 40 per cent of its labour force in agriculture. But other factors in that country, such as limited amounts of FDI, have combined with shortcomings in infrastructure and the quality of education to hold back productivity growth. Finally, the presence of a very large informal economy – a point that we return to in detail in Chapter 6 – has been a major factor limiting productivity increase. In East and Southeast Asia where productivity growth has been the strongest, it is clear that investment in human capital through better education, along with large inflows of FDI and associated technology transfer as part of a broader investment push, have all been important contributory factors.

---

[20] World Bank (2020d).

2.3.5   Human Capital

Modern theories of economic growth have placed considerable emphasis on human capital and education. In this, Asia has not only made significant progress but there is space for future gains in not only the coverage but also the quality of education. Average years of schooling have continued to rise – for example, from 2.9 to 6.4 years between 1980 and 2017 in India, from 4.4 to 7.8 years in China and from 3.7 to 8 years in Indonesia – but these are all significantly below South Korea, where the average surpasses twelve years, comparable to other high-income economies.

Although it is a useful indicator, years of schooling are not really informative about the quality of that education. For that, we need to turn to the OECD's PISA[21] which uses a common methodology to evaluate reading, mathematics and science performance for students aged fifteen years across a large number of countries. The most recent results are for 2018. South Korea and Singapore both rank in the top ten for all three disciplines, at equivalent or with higher scores than major rich world countries, such as Canada, Japan, the United Kingdom or the United States. Furthermore, some richer parts of China – the cities of Beijing, Shanghai, Jiangau and Zhejiang, Hong Kong and Macau – have reached equivalent levels. Indeed, in these four large Chinese cities, with a combined population of over 180 million, students outperformed others from seventy-eight other education systems, particularly in mathematics and science. Strikingly, the 10 per cent most disadvantaged students in these cities still had better reading skills than those of the average OECD student. Clearly, these scores would not be replicated in other parts of China, but they do testify to an impressive level of competence. These scores are not available for the Indian subcontinent. However, in Southeast Asia, Malaysia, Thailand, Indonesia and Philippines have significantly

---

[21] OECD (2020), Programme for International Student Assessment (PISA) database.

lower scores, well below the OECD average, particularly in the case of the latter two countries.

Improved access to tertiary education is also being given greater priority, either by raising the number of students at domestic universities or by sending significant numbers to study at overseas universities; a policy that was explicitly pursued as government policy by South Korea and, more recently, by China. In South Asia a similar process has been underway but largely funded privately.

2.3.6   Asia's Growth and the Role of the State

That Asia has grown dramatically is beyond dispute. That this has involved the large-scale transfer of relatively unskilled workers out of low-productivity agriculture to higher productivity, mostly urban, industry is also clear. Many countries – particularly in East Asia – have also relied heavily on export-led growth and increasing integration into GVCs. It has also been argued that Asia's growth has been built on a different approach to the role of government than that which has predominated in the now advanced economies of Europe and North America.

Emanating from post-war Japan, radiating out to Korea and then spreading eastwards again, the state has been placed far more centre-stage, whether as coordinator, financier or implementor of projects and policy. Through industrial policy and interventionist institutional and political arrangements, governments have guided the allocation of resources and provided support to selected industries in the pursuit of growth and international competitiveness. This has sometimes been referred to as the developmental state. There are disagreements about how effective such industrial policy has actually been. South Korea, Taiwan and Singapore are commonly cited as examples of good practice but there are countless cases where public interventions have favoured unviable projects or simply used protection and subsidies to support particular industries or projects. More generally, predatory behaviour involving the extraction of revenues or rents by state officials and connected private actors has been widespread across the whole region. One of the key mechanisms for rent extraction has been the ownership or control by the state of profitable firms. However, in the last twenty years this mechanism has tended to give way to other channels of extraction that do not require explicit state ownership. Influence and connections running between politicians and private businesses – as we shall see throughout this book – have become more pervasive and entrenched.

In sum, Asia has pursued multiple growth models ranging from the domestic demand-driven approaches largely preferred in South Asia to the externally oriented models of East Asia. Particularly between 1950 and 1970 governments throughout Asia adopted a more activist role that has subsequently been subject to revision. Of the mostly mixed models that emerged, none have exhausted their potential but each face substantive challenges. Common to all is the fact that significant changes in both policy and operating environments are required. What form those changes should take and how feasible they are likely to be is, of course, much of the focus of this book.

2.4 CHINA'S GROWTH: SOMETHING SPECIAL?

Although Asia's ascent has been pretty general, China's experience stands out, not least for the speed of ascent. Given that country's extraordinary success over the past decades, it is worth spending a bit of time trying to understand how that has been achieved and the wider economic consequences of the model that has been applied.

Broadly speaking, China's recent economic history can be split into the pre-1978 period when central planning, SOEs and directed spending were the main features and post-1978 when important elements of a market economy were permitted, including very large-scale inflows of FDI. At the same time, the country made a concerted push into international trade, later joining the World Trade Organization (WTO), and driving up exports, initially in low value-added activity. In 2000, exports accounted for a fifth of China's GDP but by 2007 this had accelerated dramatically to over 35 per cent. Subsequently, this share has fallen back to below 2000 levels as a result of several factors, among them the appreciation of the exchange rate and rising labour costs. But it also reflects the way in which Chinese companies have become more integrated into GVCs and the associated decline in the share of low value-added exports. In fact, China's exports have begun more closely to resemble those of high-income economies, reflecting an explicit policy focus to upgrade the quality and value of its exports while also rebalancing towards a larger services sector at home.

Domestically, since the early 1980s, the previous emphasis on public ownership has been augmented by more mixed forms of ownership and the use of different levers of policy. The state sector has certainly not disappeared, but it did shrink sharply. Rather than the state dominating all the commanding heights of the economy directly, it moved to exerting its influence and direction in more nuanced ways, including through privately-held companies who remained, nevertheless, highly susceptible to state influence and suasion (features that we will discuss at greater length in Chapter 4). The instruments of influence and control have included the pricing of energy and other inputs, including the use of subsidies, explicit and implicit. In addition, access to cheap finance and assets, along with markets and technology, has been provided to preferred companies. In fact, it is really no exaggeration to state that support to companies from the various layers of government has been pervasive in both export-oriented and domestic activities. Connected players have – particularly over the last couple of decades – been the main beneficiaries of a Chinese capitalism that has mutated but is still directed by the Party and state.

China's growth model has mainly been viewed outside the country as being driven by its focus on boosting exports. In doing so, however, the model has relied on a very particular financing strategy, the most notable feature of which has been the restraint of domestic consumption – and with it wage levels for its workers – along with a parallel addiction to investment. Household consumption has consistently fallen below 40 per cent of GDP. To put this in context, in most other emerging markets that share is between 55 and 60 per cent, which rises to 65–70 per cent in the rich economies. The flip side is that the share of investment in Chinese GDP has exceeded 40 per cent of GDP in every year since 2004,[22] while investment's contribution to growth exceeded consumption in nearly half of the years since 2000.

What has this investment boom involved? The answer is that it has mostly centred on infrastructure projects including truly vast amounts of investment in housing.[23] The massive economic stimulus launched after the financial crisis of 2008 led to a spiralling of debt. Between 2000 and 2019 debt/GDP jumped from nearly 150 per cent to over 308 per cent with much of that increase being among non-financial corporations, some of them publicly owned. Local governments also increased their exposure dramatically.

At the root of China's investment-led growth path lay a combination of very high savings rates (even now savings are around 46 per cent of GDP) coupled with punitively low interest rates on savings in the public sector dominated banking system (economists often term this sort of intervention in interest rate setting as financial repression). This has duly depressed the cost of capital, thereby permitting high investment rates alongside a dramatic acceleration in exports. All of this has at the same time imposed a massive transfer away from savers, households and, most generally, Chinese workers over decades.[24] One indicator of this is, of course, the increase in inequality. (This taxation of households and the resulting impact on consumption is a modern-day mnemonic of the sort of policies that had been used in the Soviet Union to mobilise savings to fund industrialisation.)

---

[22] Consequently, the incremental capital output ratio has climbed from around 4.2 to 6.6 between 2004 and 2019.

[23] In 1996, property investment accounted for only 2.5 per cent of GDP. By 2017 it had reached over 17 per cent.

[24] Klein and Pettis (2020) argue that this approach forms one part of a class war; the other component being the impact on workers in the rich world on account of China's external trade flourishing.

The high investment/low consumption model has increasingly been questioned, not least by Chinese policymakers. Since 2015–16, there has been greater talk of the need to rebalance the economy and boost consumption and, indeed, some action. Real wages have been rising quite fast, while depreciating the currency has no longer been an option, not least because of pressure from China's trading partners. Yet, just to shift household consumption to 50 per cent of GDP over the coming ten years, consumption growth would need to outpace GDP growth by more than 4 percentage points annually. As pre-COVID-19 annual GDP growth had fallen to around 6 per cent, this would imply a growth in consumption of 10–11 per cent per annum.

Boosting consumption also requires turning off the spigot of investment spending – to which many parts of the system are addicted – as well as addressing high savings rates along with the returns to savers. Although there has been some relaxation in the extent of financial repression, addressing the reasons for why savings rates are so high in the first place – the chronic lack of adequate social insurance, including the absence of assistance to workers falling into unemployment – has not been seriously attempted. Moreover, there are few signs that they will be anytime soon. Why that is the case we will discuss in more detail in Chapter 6.

The COVID-19 crisis has highlighted some of the strengths and weaknesses of the Chinese politics and economics. Of course, the virus originated in Wuhan, and there was a period when the authorities suppressed information that might perhaps have initially slowed the pace of diffusion of the infection. However, once the government acknowledged the outbreak fully, China's reaction was speedy and effective. As a result, output drops were relatively short-lived, and economy largely rebounded. The pandemic also accelerated the implementation of new technologies for communication, but also the application of robots and AI at work and in leisure. By the third quarter of 2020, China had restored all the output lost to the national lockdown and was once again growing, although at a lower rate of around 3 per cent. This performance compares favourably with the

United States and most of Europe, where output levels remained considerably below pre-COVID-19 pandemic levels as the second wave broke in autumn 2020.

2.5 SOME IMPLICATIONS OF ASIAN GROWTH

2.5.1 *The Depression of Poverty and the Uplifting of Spirits*

There is no doubt that growth has been accompanied by large and positive effects on other salient economic metrics. Poverty has declined in a substantial way throughout Asia. Using the World Bank's measure of extreme poverty – $1.90 PPP per day – China, Thailand and Malaysia now have poverty rates below 1 per cent, while the Philippines and Indonesia have around 6–7 per cent. Vietnam has seen a particularly dramatic and rapid decline in poverty from nearly 40 per cent in 2000 to just over 2 per cent today. Put in a longer perspective, for East Asia as a whole, nearly 1 billion people lived in extreme poverty in 1990. By 2018 this was below 30 million. In South Asia, extreme poverty remains more prevalent but has also declined quite rapidly. In India, it is estimated that around 13 per cent are in extreme poverty, while in Pakistan and Bangladesh, the shares are around 4 per cent and 15 per cent, respectively.

The $1.90 PPP measure should be viewed as a very low level of income. If using a significantly higher income threshold – $5.50 PPP per day – a level of income below which it has been estimated that someone would not be economically secure – the numbers in poverty and their shares balloon. For example, in East Asia, this higher threshold would add over 500 million to the 30 million in extreme poverty and that would be equivalent to a poverty share of nearly 35 per cent.[25] (If $3.20 PPP was applied that share would fall to 12.5 per cent). In South Asia, over 81 per cent would fall under this count (49 per cent if $3.20 was used). In other words, despite the undoubted fall in poverty that has happened in recent decades, there are still a

---

[25] World Bank (2018b).

substantial number of people who get by on low and precarious levels of income.

In short, rising levels of income have also translated into enormous improvements in living standards and, commonly, broader indicators of well-being. Average life expectancy now ranges between 65 and 75 years in India and China, respectively. A similar picture holds in other countries in the wider region. To put this in context, this is comparable to Latin America and greatly superior to most countries in Africa. However, when putting together income and non-income measures – as is done by the UNDP's Human Development Index with Life Expectancy, Schooling and Income – although South Korea sits at 22 in the global country ranking, in South Asia the main countries – with the exception of Sri Lanka – all rank lower than 130, while in Southeast Asia, many are also ranked in the 100s. Malaysia is the best performing by this measure (57) and China and Thailand still rank in the 80s.

While indicators of well-being and economic performance are enormously valuable, it has increasingly been argued more subjective measures, such as happiness or satisfaction, are possibly better indicators of welfare and certainly provide important insight into how citizens evaluate their lives and position in society more broadly. For example, self-reported happiness across a large number of countries shows that in Asia there is some, but not particularly tight, correlation between growth and happiness. That association is also – as might be expected – clearly not linear. In other words, higher incomes do not systematically and uniformly deliver higher degrees of happiness or satisfaction. There is also a considerable amount of variation both across and within countries. For example, asking citizens where they place themselves on a ladder of 0–10 where step 10 represents the best possible life, it turns out that on average Indians and Bangladeshis place themselves slightly below the fourth step whilst in China and Southeast Asia they mostly place themselves between the fifth and sixth steps. To put this in context, the countries with the highest reported happiness – for example, Norway and Australia – place themselves above the seventh step. Other data series that allow us to observe how peoples' views of themselves change over time – such as the information provided by the World Values Survey – also show that between 80 and >90 per cent of Asians report either feeling very or rather happy and that the trend has mostly been rising. To put this also in context, these percentages are not hugely different from those reported for the rich countries of Europe.[26]

If surveys of citizens' opinions tend to demonstrate a reasonably common and broadly optimistic view of their situations, questions regarding the level of trust in institutions – such as Parliament, judiciary or political parties – reveal far more divergence. For our purposes, what is particularly interesting is to gauge what people think about competition, the factors behind individual success or failure as well as attitudes towards inequality. The World Values Survey again shows that there is a great deal of variation in responses across citizens within and across countries. For example, between 40 and 60 per cent of respondents are strongly in favour of competition, including in China and Vietnam, but in both South Korea and Thailand this share drops to around a quarter.

Interestingly, although a clear majority of Asian citizens attribute success to hard work, in South Korea this falls to 16 per cent and in India this is a view held by no more than a third. A minority – ranging between 10 and 22 per cent – attribute success more to connections than hard work. We argue, of course, that connections are immensely important in the economic, financial and political worlds of these countries.

What these responses might suggest is that such connections are not – as yet – viewed as being that significant or pervasive. One interpretation of this is that the high tides of growth that have lapped these many shores have delivered enough improvement in living standards – as reflected in levels of satisfaction – to overlay any underlying concerns with how those benefits are being distributed.

[26] Ortiz-Ospina and Roser (2017).

In that regard, we might expect attitudes to income inequality to give some sense of any possible, underlying tensions. For sure, between a half and two-thirds of Bangladeshis and Indians, respectively, feel that incomes should be more equal, as do roughly a fifth to a quarter of Filipinos, Thais and Chinese. This share falls to 5 per cent in South Korea. Again, this likely points to the way in which to date the broader benefits of Asia's resurgence have continued to dominate the distributional consequences of that ascent. We now ask directly how have the benefits of Asia's large cumulations of growth been distributed?

2.5.2   *Another Side of Growth: Inequality*

It is no surprise that Asian inequality has risen substantially in recent decades. After all, this was precisely the path that the economist, Simon Kuznets, predicted in his book published in 1955. As economies develop, he argued that inequality would rise, only then to fall once they had reached high levels of income. As such, the relationship between inequality and income was expected to have an inverse-U shape. This shape has been disturbed in recent times but mostly by the fact that in high-income economies inequality has begun to rise and, sometimes, by significant magnitudes.[27] In the emerging economy world, inequality has surely increased substantially, much as Kuznets predicted.

Part of the scale of increase can also be attributed to the abandonment of explicitly socialist models of development, whether post-Mao in China or post-Nehru in India. Even so, the rise in income inequality has been quite sharp. Using the most common measure of inequality, Gini coefficients for income for South Asia now range between 0.3 and 0.4, while in China and much of Southeast Asia the Gini ranges between 0.4 and 0.5. Figure 2.6 shows the median Gini coefficient in each decade between 1980 and 2019. We see that some countries have had very high levels of inequality throughout – for

---

[27] Piketty (2014); Milanović (2016).

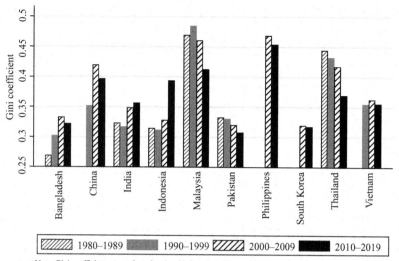

Note: Gini coefficient ranges from 0 to 1, with 0 representing perfect equality and 1 representing perfect inequality. Graph reports median Gini coefficent of each country, by decade.
Data source: Gini index (World Bank estimate)- World Development Indicators (WDI)

FIGURE 2.6 Gini coefficient in Asia, 1980–2019

example Malaysia, Philippines and Thailand – although those levels declined between 2000 and 2019. Other countries, with moderate levels of inequality, such as Pakistan, South Korea and Vietnam, show little change or a small decline over time. However, especially in the big and fast-growing economies of India and China, but also in Bangladesh and Indonesia, the levels of income inequality have been very clearly on the rise, although perhaps more so before 2000 than in the past decade.

We need to be cautious, however, when talking about income inequality. That is because its measurement remains far from straightforward. For instance, Chinese inequality undoubtedly picked up sharply after the mid-1970s with the Gini coefficient for disposable per capita incomes going up from around 0.3 to near on 0.48 by the early 2000s. From that point – according to official data – inequality has remained roughly stable. Yet, this is not very credible. Although there has been a gradual increase in workers' wages which may have helped offset any rise in the Gini, other factors – such as opening more sectors to private firms as well as pervasive corruption – could have been expected to act in the opposite direction.

In addition, wealth inequality – always far harder to measure – appears to have soared. Rapid economic growth has generated a dramatic increase in the number of very wealthy individuals, measured for example by the number of billionaires. The associated spike in wealth inequality will always be felt more keenly in countries where a significant part of the population lacks sufficient income to ensure economic security. After all, nothing could be more starkly contrasting than the ostentation of the billionaire Ambani family's vast multi-storey dwelling – equipped with helipad and whole floors for cars and employees – in the heart of Mumbai. Indeed, in the coming chapters, we will argue that rapidly growing wealth inequality can be partly traced to changes in economic policy and the economic opportunities associated with trade but much of it is also closely linked to the ways in which the connections world secures and distributes assets, privileges and revenues.

2.5.3   Institutions

Although it is undoubtedly helpful to understand the extent to which growth is being driven by factor accumulation and/or productivity improvements, relying solely on accounting exercises has serious limitations. Among them is the fact that they say little about what is wrapped up in TFP and the complex ways in which institutions and other features – including the political – affect economic performance. The economic historian Douglass North won the Nobel Prize for his work that brought institutions to the forefront of such analysis. He argued that economies which have effective institutions are able to reduce the costs and risks of transactions, and thereby enjoy higher rates of growth. Institutions were defined as the 'humanly devised constraints that structure political, economic and social interaction.'[28] He went on to distinguish two categories of institution.

---

[28] North (1990; 1991, p. 97).

The first are *formal institutions*: the laws and structures that set out an economy's economic incentives. For example, the clarity and enforcement of property right arrangements or the 'rule of law'. Other examples might include arrangements for new firm entry and bankruptcy, and for the operation of capital markets, minority shareholder protection and transparency requirements, and systems of corporate governance. The second are *informal institutions*. These are the social arrangements and norms that affect the actual (as against notional) operations of formal institutions. They include cultural factors like religious norms but also pernicious influences, such as corruption. To understand this distinction, just consider the legal system. Some countries, for example India, have well-crafted legal systems. However, problems arise because informal networks can undermine the operation and enforcement of those laws.[29] Those networks can also, conversely, help resolve the limitations of formal legal systems – for example, the time taken to get judgements – by enabling other mechanisms for resolving disputes. In countries such as Vietnam or, historically, China, the legal systems themselves have considerable weaknesses, with key concepts like private property or bankruptcy being inadequately specified. Again, informal networks can ensure that voluntary contracts will still be legally enforced. Further, although institutions can be improved – and indeed tend to be improved – as economies become more developed, the effectiveness of such changes will hinge on how informal institutions operate, particularly as all the evidence suggests that they are very likely to be highly persistent.[30]

In general, however, institutions are weaker in emerging economies than in advanced ones. Indeed, when comparing emerging with advanced economies, some have used the term 'institutional voids' to describe the situation in the latter.[31] The idea here is that when institutions are weak, transactions costs will be high and the business

---

[29] Estrin and Prevezer (2011).   [30] Williamson (2000).
[31] Khanna and Palepu (2010).

environment will be uncertain. This prevents the emergence of standards and intermediaries that can usually be taken for granted in advanced economies. Examples of such voids include accounting standards and practices; product standards and quality assurance; certification, including of professional skills; provision of credit information and the quality of the judiciary and legal system. Their absence or insufficiency holds back economic development in general, but especially of activities reliant on the enforcement of complex contracts such as in specialised markets for capital, skilled labour and technological products.

But there is also substantial variation in the form and effectiveness of institutions across emerging economies, including in Asia. In an influential book, Acemoglu and Robinson (2012) distinguished between 'inclusive' versus 'extractive' institutions. Extractive institutions are characterised by the dominance of narrow elites with economic institutions primarily designed to enrich and perpetuate the elite. In contrast, inclusive institutions tend to be based on political and economic competition and access based on the rule of law, secure property rights along, inter alia, with free media and a corpus of empowered citizens. Such broad differences clearly exist in Asia, as later chapters will highlight.

How do the economies of Asia compare with each other and the rest of the world applying standard indicators for the quality of institutions? Starting with a measure of formal institutions – the rule of law – Figure 2.7 shows where the Asian economies stand, once again relative to a regression line controlling for the level of development using GDP per capita for 178 countries in 2019. As would be expected, the quality of institutions is positively associated with the level of development, although, of course, the direction of causality runs in both directions. There is also a broadly positive association for the Asian economies, even as there is also a high level of variation among them. Some relatively poor economies such as India and Vietnam do relatively well in terms of the rule of law, while others – such as Pakistan and Bangladesh – do rather worse. In terms of global norms,

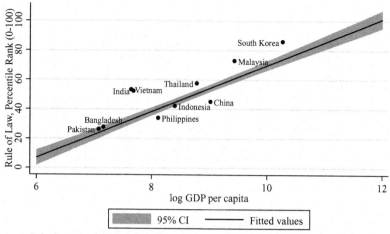

Rule of Law measures the extent to which agents have confidence in and abide by the rules of society such as quality of contract enforcement, property rights, the police, and the courts, and the likelihood of crime and violence.
Data source: Rule of Law, Percentile Rank (0-100)- Worldwide Governance Indicators (WGI), 2020 Update, Real GDP per capita in 2010 US$- World Development Indicators (WDI)
Number of countries: 178

FIGURE 2.7 Rule of law versus log GDP per capita, 2019

South Korea, Malaysia, Thailand, Vietnam and India are well above the regression line, Pakistan and Bangladesh sit on it and China, Indonesia and the Philippines fall significantly below the line.

Finally, turning to perhaps an important indicator of informal institutions – corruption – Figure 2.8 reports an index measuring political corruption between 1980 and 2019. Forty years ago, corruption was rife in Asia. That was particularly the case in Bangladesh, Indonesia, Pakistan, Philippines, South Korea and Thailand. Corruption was less present in China, India, Malaysia and Vietnam. The direction of change has also been different across the region. South Korea's rapid economic development has been associated with a dramatic fall in corruption, paralleling improvements in the rule of law since the 1990s. The situation in the other countries with high levels of corruption in 1980 – Bangladesh, Indonesia, Pakistan and Thailand – has remained more or less the same or improved/deteriorated only slightly. Worryingly, we also see a marked increase in terms of corruption in China, India and, especially, Vietnam. In Malaysia,

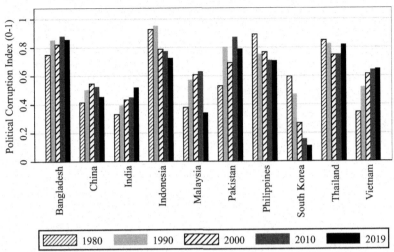

Note: Political Corruption Index scales from low to high (0-1). The index is the aggregate of public sector corruption index, executive corruption index, the indicator for legislative corruption and the indicator for judicial corruption
Data source: V-Dem [Country–Year] Dataset version 10

FIGURE 2.8 Political corruption index, 1980–2019

corruption increased until 2010 but has since declined.[32] What this suggests is that a picture of fairly good and improving formal institutions has consequently to be set alongside measures of informal institutions, such as corruption, that have shown sometimes quite divergent paths. And what is quite evident is that any assumption that growth in income levels has necessarily been associated with improvements in institutions is simply not warranted.

2.6   CONCLUSION

In this chapter we have chronicled the remarkable resurgence of Asia over the past half-century. That resurgence has translated into

[32] Corruption is of course notoriously hard to measure both conceptually and empirically, but other indicators tell a broadly similar story. For example, Transparency International in its Corruption Perceptions Index for 2019 finds corruption particularly entrenched in Bangladesh, Pakistan, Philippines, Thailand and Vietnam and much less so in South Korea and Malaysia. However, it finds China, Indonesia and India at a similar intermediate level; a more negative view of India and China than in other measures, such as V-Dem.

tangible improvements in living standards and citizens' feelings about the state of their countries more widely. But it has not by any means ensured that many of the bedrocks of wealthy economies – such as well-functioning and honest institutions – have been established. Indeed, the contrast between, and mutual dependence of, formal and informal institutions is not just a sophistic matter. What we argue throughout this book is that informal institutions and the networks of connections that lie behind them in Asia are of enormous salience and, in many instances, are the loci where decisions and actions of immense strategic importance, politically as well as economically, are taken.

The following chapters turn explicitly to the task of showing that the ubiquitous network of connections that can substitute for market-based transactions when institutional structures are insufficiently developed should also be viewed as powerful informal institutions. Further, these informal institutions have proven highly resistant to change and to replacement by more formal, and more transparent, institutional arrangements. We now turn in Chapter 3 to looking directly at the ways in which they are configured and, hence, to the contours and composition of the connections world.

3 The Power of Networks of Connections

The connections world is a pervasive feature of modern Asia. But what does this actually mean? How are connections formed and how do they play out? This chapter spells out in detail what we mean by the connections world and describes how it functions across Asia. The principal dimension is the way in which businesses can leverage their connections to politicians and politicians to businesses for their mutual benefit. In so doing, connections are rarely just bilateral, that is, involving connections between only two persons or entities. Rather, they mostly have strong network dimensions which link up a wide variety of players and even materially affect how businesses configure themselves. These more complex set of relationships ensure that the ties that bind politicians and political parties to businesses and other individuals are not only economically and financially meaningful, often reciprocal, but also rather persistent. We start the chapter by describing the sorts of networks that exist in Asia before then giving colour and detail to their functioning by drawing on a wide range of examples and episodes.

Most discussion of Asia's present and likely future has concentrated, understandably, on the political systems and institutions that have been associated with its remarkable economic renaissance. Beneath this general ascent can be found no singular confection of political system but rather a wide mix. Yet a focus on the differences – and relative merits – across political systems risks obscuring what is indeed a fundamental – and surprisingly common – feature of all Asian countries. That feature is the role played by the connections world where webs of ties formed through networks create and facilitate privileged access and rewards in ways that have proven to be highly resilient. This world cuts across political systems – autocratic and democratic – with surprisingly common *modi operandi* and consequences. As such, it provides the backbone of systems that appear so very different when looked at through the lenses of political systems or institutions.

Networks are, of course, ubiquitous in human life and exist for many purposes as they describe the web of social relationships that surround individuals. Their formats consequently vary widely and encompass groups of friends and families, university alumni, chambers of commerce, sports fan clubs and religious groups, among others. Between any groups of people, there are a large number of possible connections or ties, but in practice relatively few people are actually connected. The proportion of people in a network who are connected is measured in terms of density. At a wedding, density will likely be high but on a train, low. Some networks are configured in a dense way, others are more diffuse. Moreover, within networks, some ties are weak and some strong.[1] How they are configured often reflects the purposes to which they are being put. For example, dense networks tend to be effective for leveraging assets and ensuring cooperation. Diffuse networks tend to be effective in providing access to, and transmitting, information. In general, the idea of having access to a network and networking is thought to be not only acceptable but almost natural. Being part of some networks is seen as being highly desirable – the World Economic Forum, with its annual Davos meeting, is a prominent example. In politics, family networks and ties are common phenomena. Belonging to such a network and the place of a family or families in a network – particularly how centrally located they are – has been found to be important for getting elected in many diverse contexts, whether village or municipal elections in the Philippines[2] or the leadership of India's Congress Party. Yet the power of networks can have a far darker side – mafia, criminal organisations

---

[1] Granovetter (1973).   [2] Cruz et al. (2017).

and drug cartels are all forms of network organisations whose structures are put together in ways that help bind members and facilitate criminality.

Less dramatically, but often with more effect, networks can be leveraged to limit competition or provide preferential and discretionary access to sources of value whether of assets or revenue flows, including in ways that are tantamount to theft. In earlier times – medieval and early modern Europe, for example – non-agricultural economic and political life was mainly organised around economic associations or guilds; forms of network. In England, the guilds comprised craftsmen such as goldsmiths or masons and had been established by Charter from the King or City, being typically granted a monopoly on their trade within the relevant jurisdiction.[3] A key feature of the guilds system was therefore its ability to limit competition by both restricting supply and regulating entry into their professions. The guild system remained alive in Europe well into the nineteenth century, with vestigial traces surviving even now.

In this chapter, we argue that the networks connecting politicians and businesses in Asia have particular prominence. Such networks seek to derive advantage from their mutual – often reciprocal – relationships. Further, there are also network dimensions to the very ways in which both political entities (obviously) and businesses (less obviously) are organised and function. Indeed, their dynamic interplay summarises much of the connections world.

We first describe and map the networks of businesses and politics that exist in Asia, showing that there are differences in the way networks are configured. This is due to a variety of factors, including the size of the country, but also to differences in the broader political systems. The most striking difference is between the network structure and ways of working in China, as against other countries in Southeast and South Asia. The intricacies of how networks are leveraged and their consequences take up the second part of the chapter

---

[3] Braudel (1992).

where a series of examples are deployed. We use cases and episodes not just because they provide an effective way of showing how networks and their connections actually function but also because, by their nature, many of these connections operate below the surface or are even actively camouflaged. This naturally results in limiting the availability of standardised and large datasets. The cases that we use – despite their idiosyncrasies – bring out in sharp relief many common features, not least with respect to their purpose, but also in the ways in which businesses are organised and function, a theme we return to in Chapter 4.

3.2 DIMENSIONS OF ASIAN NETWORKS

In Asia, capitalism has spread its tentacles everywhere, even if some regimes have political foundations that pay obeisance to socialism. All systems, however, still remain massively dependent on informal institutions and relationships, many of the most significant being of a network nature. Whilst formal institutions may affect the rules of the game, such as laws and rights, informal institutions commonly work through cultural norms.[4] The result is a pervasive coexistence of formal and informal institutions and their associated ways of behaviour. This coexistence has caused its own tensions. For example, the interaction of financial instruments, created for a world where formal institutions are dominant and arms-length transactions are the norm but deployed in a world where relationships matter more, can be explosive. The resulting risky behaviour culminating in the 1997–98 East Asian crisis is a vivid case in point.[5]

Irrespective of the variety of Asian capitalism, there are very porous boundaries between politicians and business and between firms and the state. Many successful businesspeople go into politics and politicians often have stakes – frequently opaque – in commercial activities. Recent cases that, for one reason or another, have attracted public scrutiny or reached the courts, whether in India, Malaysia or

[4] North (1991).   [5] Rajan and Zingales (2005).

Thailand, indicate that those stakes can be very substantial. Others use the authority of office to set up family members in businesses, sometimes supported by lucrative state contracts and always aided by the contacts and influence of their promoter. Although most Asian countries do require officials and elected politicians to declare their assets, including real estate and shares, along with income and its sources, there is a large discrepancy between what is notionally mandated and actual compliance, let alone penalisation for non-compliance. Similarly, although companies are mostly allowed to make transparent donations with explicit caps to political parties or candidates, the opportunities for covert donations are legion and compliance, once again, is weak.

This world of porous boundaries is sometimes collapsed to the concept of cronyism, where cronyism is defined as the receipt of contracts, office and/or other privileges simply on the basis of some form of connection. Such opportunities present themselves particularly when there are ambiguities over control rights, as in contemporary China. But cronyism is describing a rather specific form of privilege, namely the appropriation of revenues or assets or both by private individuals from institutions that are meant to be publicly owned. Yet, networks of connections are commonly deployed in ways that go far beyond the leveraging of public assets and organisations and have much broader significance including, as we will argue later, for even the sustainability of economic development. Further, the perpetuation – and resilience – of networks is not simply inertial, far from it. Networks comprising individuals, politicians, political parties and companies – public and private – tend to be highly adaptive and purposive. So, although the mechanics of operation may vary significantly country-by-country, the underlying motivations tend to be very similar.

The instruments used to achieve particular objectives vary. As we noted in Chapter 1, politicians commonly look to companies to make campaign or personal contributions (over or below the table), pay bribes, provide reciprocal favours – such as creating jobs in regions or times that are politically advantageous or making discretionary donations to causes favoured by those in power – and, in some instances, execute important dimensions of public policy. An interesting example of a traditional exchange of favours in a contemporary setting is firms' use of their corporate social responsibility budgets to support causes high on the agenda of politicians. This has happened in South Korea and more recently in the Philippines and Indonesia around the COVID-19 pandemic.[6] At the same time, companies variously look to politicians to grant them privileges whether through public contracts, tax breaks, access to assets, investment resources, subsidies, formal financing (as from state-owned banks) and help in securing markets, as well as advantages that might be delivered through skewed or absent regulation, competition rules and other policies.

Although this encompassing, transactional framework gets to the heart of what most interactions are about, it misses some of the ways in which individuals and entities actually organise themselves to achieve some of these goals. Indeed, in how they organise, there may be a strong network dimension that coheres with, or mimics, the ways in which politicians and companies interact. For example, the institutional framework for private companies, and the ways in which they may link to other companies, can well be designed with a central – even primary – purpose of leveraging resources and assets as well as gaining advantage, whether in relation to the regulator or with regard to other actual or potential competitors. This is indeed reflected in a prominent feature that differentiates Asia from most advanced economies. That is the presence of business groups or conglomerates as the dominant institutional and organisational format for companies.

What explains this predilection for conglomerates or business groups? We will discuss this at more length in the next chapter, but

---

[6] Choi et al. (2018), *Financial Times*, 'Duterte apologises to big business after help with Covid 19', 8 May 2020.

one common interpretation is that this is caused by missing markets – such as having access to capital markets – and missing institutions. In other words, it is an efficient, if second best, response to institutional lacunae. Yet this only captures part of the picture. The predilection for business groups also reflects specific responses to the institutional context, whether for funding or control. And it reflects the way in which economic power and political power interact. Adoption of the business group format consequently has much to do with the way in which such organisational forms offer advantages when dealing with politicians as well as with large family structures. Such organisational formats facilitate hedging business and politically-related risk including through opacity in accounting and transactions while providing 'better' vehicles for negotiating with politicians. For example, scale and complexity can act as good deterrents to politicians who might be tempted to try and expropriate. Business groups may also be a way of responding to predatory states.

3.3   NETWORK SHAPE AND MEASUREMENT

How these networks and the interactions they facilitate are shaped and work on the ground depends on history, political systems and income levels (the latter two being often highly correlated) along with other sociocultural features, such as inheritance rules and practices. Probably the main conditioning factor is the political system. The members of political networks – represented as nodes or points of linkage in the network – include not just politicians and political parties but also their friends, relations and SOEs along with any privately-owned companies with whom they have affiliations. Whenever nodes have connections with another, this is known as an edge or tie; denser networks have more ties.

Measuring networks is obviously far from straightforward, not least because individuals and entities – such as companies – often try and cloak their involvement, either because of corrupt behaviour or because they prefer to block others – including the public but also competitors – from getting clear sight of possible links and hence possible benefits. Despite these difficulties, in recent times there has been an international push for greater transparency. This has been driven by a combination of a growing concern about corruption and money laundering, as well as investor pressures to limit exposure to companies with flawed governance. In democracies, such as India, journalists have also increasingly delved deep to throw light on deals between connected parties in order to identify possible preferential and unfair treatment, including any implications for the public finances. Among the consequences of these combined pressures have been the collation of countrywide lists of politically-exposed persons and entities by major financial services companies. This in turn has been aided enormously by the growth in artificial intelligence that has been deployed in searching across all sorts of publicly-available databases, local and international, for connections.

For our purposes, we will use one such database that covers almost all countries in the globe and all of Asia.[7] It contains the names of politicians, political parties, individuals and companies as well as the connections that exist between these different components. The data have been collected from a wide array of sources. These include regulatory, law enforcement, sanctions and other lists as well as the specific collation of information on politically-exposed persons. As such, they give us a view of the size and configuration of the various country-level networks and allow mapping how particular individuals, politicians and businesses are connected to each other. Although it is far from comprehensive,[8] it does allow a detailed look at each country, as well as facilitating comparisons between countries. Before directly using this data, some wider comments about network differences across different political systems is in order.

---

[7] Commander and Poupakis (2020) provide a longer description of the data and provenance

[8] This dataset, along with those collected by other commercial providers such as Dow Jones, has been collected with the aim of providing information on the politically exposed and other risky individuals or entities. This sort of information is obviously useful for investors. It also allows us to quantify networks in these countries.

3.3.1 Networks and Political Systems

The composition and shape of networks varies significantly across different types of political system. In democratic systems, networks are organised around a multiplicity of institutions, such as political parties, whereas in autocratic regimes, it is more likely that this is the case around individuals or entities or a unique political party. For example, in a dictatorship a hub and spoke network structure is likely to be present, as links flow to, and from, a small number of, or even unique, central nodes. This is stylised in Figure 3.1 where the black circles are nodes and the lines between them are ties or edges. The size of the node reflects the number of ties or connections that they might have. The figure shows a central and large node with a few other far smaller nodes with their limited number of ties.

Very few political systems are actually as simplified, of course. That is because autocracies themselves are mostly quite complex constructions. Even so, political networks will tend to reflect the power structure. In autocracies this often takes the format of a

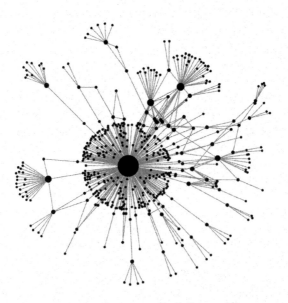

FIGURE 3.1 Network types: stylised dictatorship

pyramid under the ruler followed by any supporting coalition of actors along with any opposition that is tolerated. Moreover, the extent of integration of the network in most autocracies – as measured by the share of nodes and ties contained in the largest component of the network (sometimes known as the Big Island) – tends to be quite low.[9] This is attributable to the divide and rule strategies applied by autocrats whose aim is to limit the opportunities for others to combine and act against them and who purposefully impede the formation of networks. Exceptions to this are the so-called socialist systems operating in China and Vietnam. In these places – and China, in particular – the existence of the singular political – Communist – party ensures that the network connecting the politicians, party and business is actually quite highly integrated. For example, representatives of the Chinese Communist Party play a governance role in most large organisations, including firms and universities. The formation of opposition networks is limited by other –often repressive – means. The constellations of rivalries that are mirrored in networks are, in these instances, largely played out within the monolithic Party structure as also, sometimes, across the various geographies or provinces. It is very clear that the different shades of autocracy are reflected in corresponding differences in the structure and, indeed, composition of their networks.

Such variation is also true for democracies. However, most nodes in their networks are organised around a multiplicity of political parties, politicians and, less commonly, state-owned companies. Moreover, they tend to have more complex and denser networks, and these are almost always far more integrated than is the case on average for autocracies.[10] In democracies, most nodes and their ties are contained in the largest network component and the network as a whole is consequently more integrated than is the case for

---

[9] In the language of network analysis, a component is defined as a subset of nodes where all its members are connected with at least one other member of that subset.

[10] Commander and Poupakis (2020).

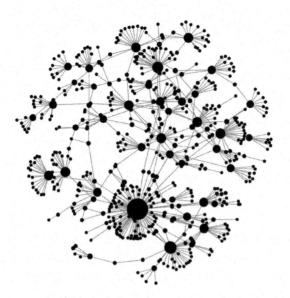

FIGURE 3.2 Network types: stylised democracy

autocracies. This is stylised in Figure 3.2 which contains a multiplicity of nodes, some of which would be political parties. These in turn have their ties to politicians as well as to companies and other individuals.

Finally, where a person or entity is placed in a network will often be of material importance in terms of their effectiveness. Different political systems, as we have hinted, offer widely varying opportunities for occupying strategic positions. Having a strategic position not only presupposes being in the main – or giant – component but also having a central location in that component. For example, that could be achieved if a person, say, lies between other nodes so that the latter tend to be dependent on that person for access to information or resources. A further indicator of the strength of a person or entity's location in a network is also to whom they are connected. In other words, if a person or company is connected to other important and strategically placed nodes, it is probably more likely that they will be able to leverage those connections in some way.[11] This may manifest itself not just within a country but also across countries, a feature touched on later in this chapter when we discuss the texture of the actual networks that exist in Asia.

3.4 FEATURES OF ASIAN NETWORKS

A good way to appreciate the differences in networks across countries is by means of network maps generated using the information from our large dataset. But before doing that, Table 3.1 summarises the main features of the networks across the Asian countries. The first thing that stands out is that although the absolute size of the networks varies substantially across countries, the differences are generally much smaller once population is accounted for. When that is done, Malaysia emerges with by far the largest number of nodes and ties followed by South Korea, Philippines and Thailand. With the exception of Bangladesh and the Philippines, most networks are quite highly integrated with substantial shares contained in the largest component. Further, in all countries, individuals and politicians make up most of the network, mostly over 90 per cent when measured by their share in the total number of nodes. But there are some exceptions. In Malaysia private firms and SOEs comprise nearly a quarter and in Indonesia private firms alone make up more than 20 per cent of the network. Finally, although our coverage is significant, by no means does it pick up the full extent to which politically-exposed persons have links to companies. For example, the degree to which they are connected through proxy shareholdings or other forms of non-transparent beneficial ownership or control over entities or cash flows is often underestimated or even missed. This can be a significant omission given the huge role of business groups in Asia.

Who occupies a strategic or central position in these networks? Self-evidently, institutions, such as large banks and other finance

---

[11] In the network literature, having a central or strategic location is often measured by betweenness while taking account of whom you are connected to is measured by eigenvector centrality.

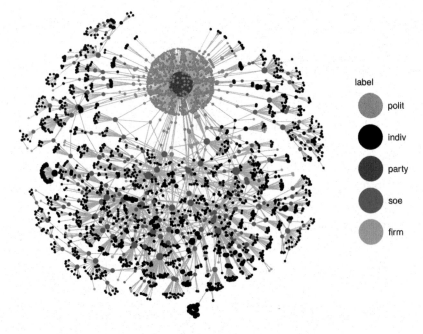

FIGURE 3.3 Network map: Vietnam

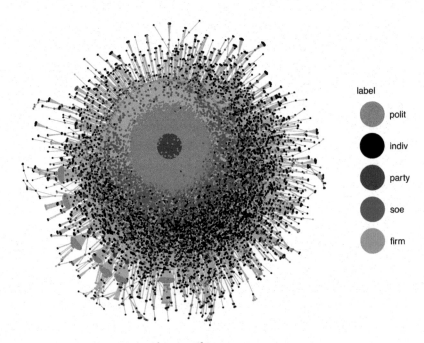

FIGURE 3.4 Network map: China

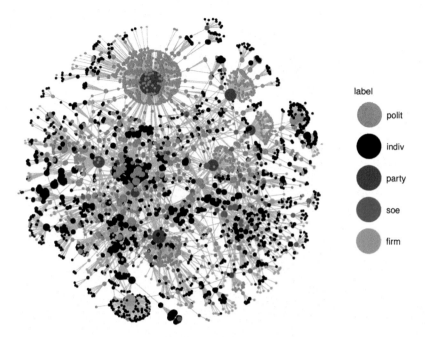

FIGURE 3.5 Network map: Philippines

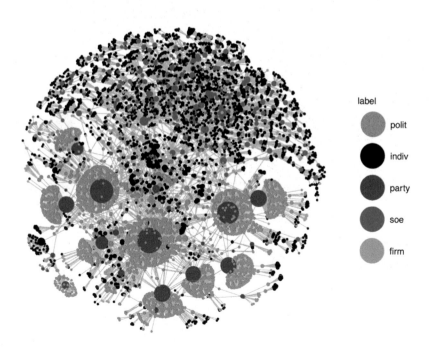

FIGURE 3.6 Network map: Indonesia

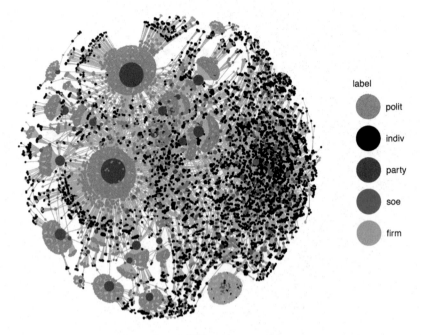

FIGURE 3.7 Network map: India

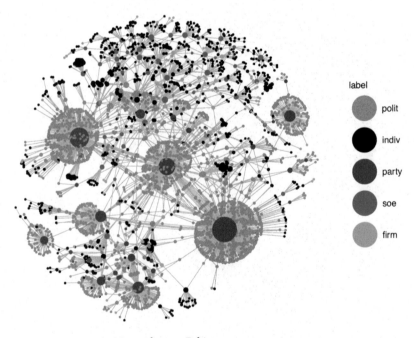

FIGURE 3.8 Network map: Pakistan

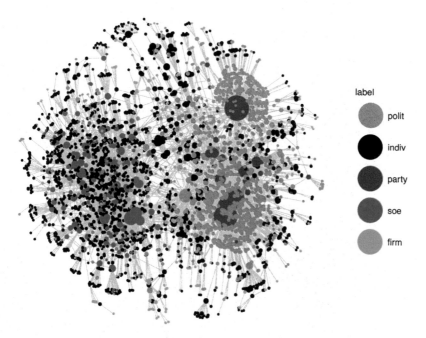

FIGURE 3.9 Network map: Thailand

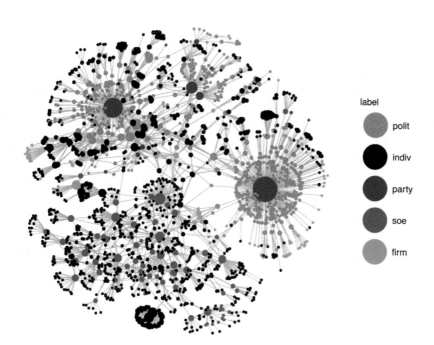

FIGURE 3.10 Network map: Bangladesh

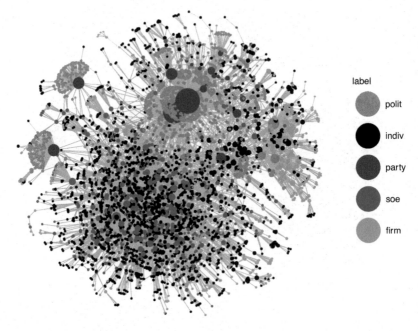

label

polit

indiv

party

soe

firm

FIGURE 3.11 Network map: Malaysia

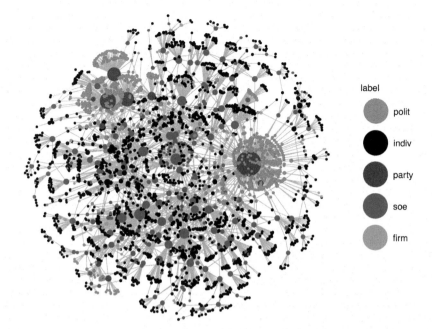

label

polit

indiv

party

soe

firm

FIGURE 3.12 Network map: South Korea

Table 3.1 Characteristics of Asian networks

	Bangladesh	India	Pakistan	Thailand	Malaysia	Philippines	Indonesia	South Korea	China	Vietnam
Political regime	Mixed	Democracy	Democracy	Mixed	Democracy	Democracy	Democracy	Democracy	Autocracy	Autocracy
Income level	Lower Middle	Lower Middle	Lower Middle	Upper Middle	Upper Middle	Lower Middle	Lower Middle	High	Upper Middle	Lower Middle
Network size										
Nodes	8,243	30,352	7,882	7,612	7,844	13,654	17,263	7,516	85,432	8,246
Edges	13,286	62,307	14,839	19,598	22,407	29,014	34,303	12,985	19,8232	16,503
Nodes per capita	0.05	0.02	0.04	0.11	0.25	0.13	0.07	0.15	0.06	0.03
Edges per capita	0.08	0.05	0.08	0.28	0.32	0.28	0.13	0.25	0.14	0.18
Integration										
Share of giant component	0.26	0.59	0.48	0.56	0.76	0.29	0.52	0.49	0.75	0.4
Composition (% of total)										
SOEs	0.01	0.01	0.01	0.03	0.09	0.01	0.02	0.05	0.08	0.05
Private firms	0.02	0.04	0.04	0.05	0.13	0.03	0.21	0.03	0.03	0.03
Individuals	0.85	0.53	0.54	0.54	0.51	0.54	0.41	0.65	0.29	0.65
Politicians	0.11	0.4	0.4	0.38	0.26	0.42	0.25	0.26	0.6	0.26

institutions, will – largely as a consequence of their function – have not only a large number of connections but these will also tend to be centrally located. In China and Vietnam, the respective Communist parties also by design occupy a central role, even if there is likely to be a great deal more complexity than might appear to be the case due to both the levels of decentralisation in the Party and the nature of links running between the Party, Party members (politicians) and other individuals and entities. But – as we shall see in the next section when discussing the actual texture of networks – much of the centrality that matters exists between a limited number of players or nodes.

With this in mind, and for each country, Table 3.2 reports a measure – betweenness – that identifies the extent to which a node lies between other nodes. This is done for each category and the columns sum to 100.[12] What emerges is that in China, most especially, the unique political party accounts for over two-fifths of total betweenness and this share is far higher than for any other country. Vietnam, by contrast, despite having a broadly similar political arrangement has a fairly low share for their unique party. Instead, the numerous state-owned enterprises account for nearly half. In the two largest South Asian countries – India and Pakistan – political parties also account for high shares. SOEs' shares are significant in most countries – Philippines and Pakistan being exceptions – and especially so in Vietnam and South Korea. Politicians, throughout, account for between a fifth and a third of the total while other individuals' shares vary significantly across countries, with a particularly low share in China.

3.4.1   Visualising Networks

Turning now to visual representations, Figures 3.3–3.12 provide country-level network maps. They have been constructed using the

---

[12] What this means is that the betweenness share is defined as the ratio of the sum of betweenness of all nodes for a particular category (e.g., SOEs) over the sum of the betweenness of all nodes in the network.

Table 3.2 *Being well placed in a network: betweenness shares*

Country	Bangladesh	Pakistan	India	Thailand	Malaysia	Indonesia	Philippines	China	Vietnam	South Korea
Betweenness shares										
Political parties	22	38	31	23	20	25	28	41	16	18
SOEs	23	13	24	27	38	28	9	26	49	45
Private firms	4	3	1	1	2	1	5	1	0	1
Politicians	26	33	29	30	23	29	40	31	21	21
Individuals	25	13	16	19	17	17	19	1	14	15
	100%	100%	100%	100%	100%	100%	100%	100%	100%	100%

database on political parties, politicians and other politically-exposed persons and entities – state-owned as well as private – that we have already mentioned. What the maps do is take each entry in the database, account for its type – whether it is a politician or company, for instance – and enter it as a node. That node is then scaled by the number of ties or edges that it has. In the case of a political party, we could expect there to be a significant number of ties, primarily, but not exclusively, with politicians. The maps allow us not only to visualise the network as a whole (or more exactly, the largest component of the entire network or database) but also to identify which are the nodes with large numbers of ties as well as showing how these ties are directed. As we shall see, these maps also allow getting a sense of the differences in shape and composition of the network, including across political systems. For instance, demonstrate the central place of the monolithic political (Communist) party in China and, to a much lesser extent, in Vietnam. In these instances, the network maps resemble a Cyclops eye, with party members (politicians) clustering around the hub or axis of the Party which then radiates out to the large numbers of SOEs. In sharp contrast to what would have been the case even thirty years ago, these networks also contain a significant number of private companies and other individuals but, for the most part, they are less central.

In the case of the Philippines and Indonesia, the former has a major political party and cluster of politicians along with a limited number of smaller parties. The network space is also populated by a significant number of other individuals along with private firms. The maps also indicate significant ties between politicians and private firms. SOEs are not very common. In Indonesia, by contrast, there are eleven political parties and clusters, three of which have a far larger number of edges or ties than the others. The network space is also occupied by many SOEs, although there are links between politicians and private companies. For India, the two main political parties – Congress and BJP – stand out but there is a multiplicity of other political groupings, many of which are regional parties. The rest of the network space is populated by a substantial number of SOEs, but also individuals and private companies. In Pakistan although there are at least nine political parties with clusters, the rest of the network is relatively lightly populated with individuals and private companies and a comparatively small number of SOEs, at least when compared with its neighbour, India. The other countries mostly have several major political parties with, however, differing compositions of individuals, private firms and SOEs in the total network. Thailand, for instance, has two main political groupings with some prominent SOEs but what is most striking is the number of private firms and other individuals with links to those parties and politicians.

3.5 TEXTURE OF NETWORKS

So far, we have described the broad format and composition of Asian networks of connections. Whilst certainly informative, these country mappings necessarily miss the detail and texture that can allow us to see how networks of connections actually function. It is obviously important to know with whom someone is linked, but meaningful questions also concern, why and what for? In other words, how do actual connections get structured and play out? For this, we will have to move to specific cases where qualitative evidence is available. When doing this, however, it is as well to remember that our network maps differ in format, and doubtless in texture, across different political regimes. Recognising this, we start the discussion by reference to those Asian countries where some variant of autocracy is in place, before moving to the more numerous democratic instances. Our approach throughout is to use more in-depth examples. Fortunately – or otherwise – there is no shortage of such narratives!

3.5.1 In the World of Autocracy

In recent times, Asia's current autocracies – notably China – have been pinpointed as epicentres of opaque connections and associated transactions. The cyclops at the heart of China's network map suggests an obvious starting point. Political systems based on one party rule have always developed parallel forms of organisation that use connections and networks to exploit or circumvent official channels and constraints. In the Soviet Union, for example, this was encapsulated by the term, *blat*. Loosely translated as 'favours', the transactional norm was often in goods and frequently corrupt. But what *blat* signified was, above all, the importance of networks outside of official channels, not just for securing scarce goods but also getting things done, including in the official economy itself.[13]

Some of the flavour of this is also echoed in modern-day China. The official monopoly of political, and sometimes economic, space occupied by the Chinese Communist Party (CCP) disguises the existence of many networks running through it, but also running alongside it. In fact, the existence of networks and the purposes to which they are put far exceeds in complexity and scope, that which developed in the interstices of the Soviet Union. That is because not only does the state co-exist with a large private sector but also because cumulative measures of economic reform implemented over past decades have still left major ambiguities over who owns and controls what. Consequently, while the mapping of China's network in shows that the apex of the system remains the CCP, there is a multiplicity of connections to and from and between private firms and individuals. The purpose of these connections, and the pathways they take, in some, but by no means all, instances clearly run counter to what the Party notionally stands for, or what public policy intends.

The role of connections in China – summarised by the widely used term, *guanxi* – is endemic. Most commentators have focused on how connections have commonly been used for corrupt purposes, including at or near the top of the political system. Such instances have proliferated, the reasons for which are ascribed – as in Minxin Pei's influential book on crony capitalism in China[14] – to a mix of decentralisation – with greater power and discretion vested in Party officials at province, county or municipal level – alongside a continuing and pervasive lack of clarity over ownership and control rights.

---

[13] Ledeneva (1998).   [14] Pei (2016).

Consequently, the substantial state sector, along with murky or loosely determined lines of demarcation and muddy property rights, continues to offer opportunities for officials and their associates to translate public resources into private ones; what economists call rent seeking. At the same time, the State and Party retains enormous power and discretion in regulatory and allocative decisions. How that power is exercised lies at the heart of far more than a set of corruption stories, not least because it provides a fulcrum around which connections are formed and put to use.

While connections can be exploited in myriad ways, most can be meaningfully placed in four main categories that are organised by a combination of purpose and composition.

The first involves the use of position and access by officials to tunnel out private gains from publicly owned assets, such as SOEs, commonly exploiting the ambiguities concerning ownership and control rights that we have already mentioned. However, this sort of activity can be both time consuming and occasionally risky. Moreover, the sorts of officials who have the ability to act in this way often lack the entrepreneurial capacity with which to capitalise fully on those tunnelled assets.[15] Nevertheless, such behaviour still persists, albeit with a smaller pool of potential resources due to the declining size and scope of the SOE sector in China.

A second channel that has become far more prominent in recent decades has been for public officials – Party members – to develop networks, often dense and persistent, of connections with businesspeople. The main instruments that are used tend to be the issuance of permits, licenses and other instruments of access, including allocating ownership or control rights to particular parties or entities on preferential terms. Those preferred parties may be family relations or associates, as well as other businesspeople to whom the official is connected. Such practices, which we consider in more detail in Chapter 4, extend right across the political system.

---

[15] Pei (2016, p. 117).

Although having explicit business ties may not be seen as acceptable for top-level leaders of the CCP, having their children or other relations build large business interests on the basis of their connections has undoubtedly been occurring. Both the current president and his predecessors' extended families have accumulated very substantial wealth and business assets.[16] A number of the public corruption trials of high-level Party officials that have occurred over the last decade – not least the infamous case of the former Member of the Politburo and Party Secretary of Chongqing, Bo Xilai, along with his wife and other officials – have highlighted how significant such behaviour has been. But this sort of collusive behaviour exists at many levels of the party machine where connections with private businesspeople may be put to the service of providing access to resources, principally land, as well as permits for exploiting natural resources. In fact, it can almost be said that in most parts of China, providing cheap land has become one of the main currencies of these connections. Many local governments have become highly dependent on revenues from land sales while also using land as collateral for borrowing.

There are numerous examples of the way in which acquisition of land rights and permits has been affected by political connections. For example, in the celebrated case of another former member of the CCP's Politburo and Standing Committee – Zhong Yongkang – his son was able to acquire rights to an oil field at a price less than 5 per cent of its actual value, while another businessman connected to him gained control over a coal mine in Yunnan, again at a hugely discounted price. Although these practices are very widespread, the benefits have been distributed very unequally. A very detailed study of land transactions by local governments over more than a decade found that firms linked to China's political elite – the Politburo –

---

[16] See, for example, the investigation of Wen Jiabao's family's wealth; www.nytimes .com/2012/10/26/business/global/family-of-wen-jiabao-holds-a-hidden-fortune-in-china.html. See also Shum (2021) for a detailed discussion of how some of that wealth was accumulated.

secured average price discounts in excess of 55 per cent compared to those without such connections. Moreover, the higher up the political tree the connection was, the greater the discount. For instance, a firm connected to the top level of the Party – the Politburo Standing Committee – received an additional discount as high as 20 per cent. The quid pro quo for these sales was quite commonly the promotion of the provincial Party secretaries in whose jurisdiction these concessions were granted.[17]

A third – and important – strand of connections runs between the State and selected private companies or business groups. While these sorts of connections may have been founded at some point on informal ties, over time they have become explicit and often a major strand of public or industrial policy. Manifestations of privilege have sometimes, as we will discuss in Chapter 5, involved explicit trade protection but often it has been simply preference that has been exerted. In software, for example, local companies benefitted by being able to flout intellectual property rights but also from substantial government purchases and contracts.[18] More recently, in e-commerce – where some major Chinese companies have emerged, including Alibaba, Tencent and Baidu – all have maintained strong connections to government and are treated – and touted – as national champions.[19] In the following chapters, we shall explore more how these connections are playing out.

The way industrial policy preferences unfold is also very well illustrated by ZTE Corporation and Huawei Technology, leaders in telecommunications manufacturing, who together account for over $210 billion in yearly revenues. ZTE is a Chinese state-holding company and Huawei is private, but both have rather similar ways of connecting with government. Already in the late 1990s, both received preferential treatment in the form of duty-free import of new

---

[17] Chen and Kung (2019) who used information on over 1 million land transactions between 2004 and 2016 in which local governments were the sole seller.

[18] Saxenian and Quan (2005).

[19] As an example, for a discussion of Alibaba, see Yu (2018).

technologies, buy-local campaigns and instruments and materials for R&D. For ZTE, even though by 2004, state ownership had fallen to below 50 per cent, other shareholders were former employees of the original state enterprise. Indeed, family of the current leadership are known to have a major interest in the company. As such, the state and these insiders count for much more ownership than legally defined.[20] This configuration is not at all uncommon in China. Thus, the public–private enterprise distinction is made opaque by China's state enterprise reform experience. Similarly, Huawei founder Ren Zhengfei, a former member of the People's Liberation Army, has had visible support and guidance from the state and the military.[21] In the founder's words, 'If there had been no government policy to protect [nationally owned companies], Huawei would no longer exist'. Ren initially followed some of Mao Zedong's agenda, such as providing services to rural customers and local governments were important in purchasing and promoting Huawei's equipment. Ren also managed to leverage his networks to win contracts for military telecom networks.

In short, both ZTE and Huawei have derived significant business from government and large state enterprise clients, such as China Unicom and China Telecom. Huawei has been estimated to have received between 2008 and 2018 as much as $75 billion in state support as part of its drive to move from making phone switches to becoming a leading, global telecoms equipment manufacturer.[22] Around 60 per cent of that support was financial – loans, credit lines, tax waivers and other low-priced finance from state-owned lenders – with a further 33 per cent due to explicit incentives provided by the Chinese government. Access to these subsidies naturally helped the company undercut competitors, including winning foreign contracts.

A fourth set of connections runs between SOEs and government. The tone and texture of these links often reflects the fact that

---

[20] Ahrens (2013).   [21] *Far Eastern Economic Review* (2000).
[22] See www.wsj.com/articles/state-support-helped-fuel-huaweis-global-rise-11577280736.

most of this relationship is organised around bargaining for resources, whether for investment or current expenditures, such as for the preservation of employment, something we will discuss in more detail in Chapter 6. These features are by no means unique to China, having been a central element in all the planned economies of the former Soviet Union as well as in Central and Eastern Europe.

3.5.2   Among Democracies

The complementarity between politics and business in democracies has been revealed repeatedly in the Asian context, sometimes taking forms that are common worldwide – donations to political parties and politicians by business being an obvious case in point – but also in ways that are rather specific to either the country, sector or type of activity. In South Korea, the recent Choi-gate scandal involved the country's major business groups – including the massive Samsung group – channelling funds to a long-time confidante of the then president (both of whom were jailed) to secure approval for business deals and a merger.[23] While clearly corrupt, the aim of such action by the companies involved was to achieve business goals that required government approval.

Direct political participation by businesspeople has also been seen as facilitating access to assets, resources or concessions and as a way of preserving accumulated interests. In many instances, a division of labour has been used with one family member participating in politics while immediate family members retain custody and management of family businesses. For example, the revolving doors between business and politics are sharply illustrated in Indonesia. Aburizal Bakrie – the unsuccessful presidential candidate for the business-oriented Golkar Party in 2009 – was, with his brothers, the owner of the Bakrie Group, a diversified business group with major interests in coal mining.[24] The next unsuccessful challenger for the presidency in

---

[23] The implications of this scandal for Samsung are discussed in www.ft.com/content/8035a965-19fb-4b6f-88ca-d301073d0e38.

[24] EIU (2016).

2014 and 2019 – Prabowo Subianto – exemplifies even more how family, political and business connections work. His father was former President Suharto's Minister of the Economy and he himself married Suharto's daughter. Despite having an army background, he accumulated significant wealth and a large group of businesses with, inter alia, interests in oil and gas, palm oil and mining. Yusuf Kalla – a two-term vice president to two different presidents – also retained active business interests through his Kalla Group. The media tycoon, Hary Tanoesoedibjo, who acquired his assets from the Suharto family went in and out of various political parties as he vied to be a vice presidential candidate before founding his own (United Indonesia Party) in 2015 while aiming to use his media platforms to support his political quest.[25] These rather broad sets of channels of interaction between business and politics belie a common underlying purpose, namely, to gain advantage and to leverage ties.

A stark – and almost uncomplicated – example of the intersection of politics and business concerns is Thailand and, specifically, the election of Thaksin Shinawatra and his Thai Rak Thai Party in 2001. Shinawatra had built his fortune in telecoms but had diversified into a wide range of activities, real estate, aviation and satellites as well as extensive trade deals with neighbouring Myanmar and China. The pattern of diversification always reflected the ability to leverage connections to secure access to rights or resources. Whilst in power, Shinawatra had transferred shares in his Shin Corporation to family members, but the company was alleged to have continued to benefit from preferential access to resources, as well as finance from a state-owned bank. Subsequent court rulings found that Shinawatra abused his position to transform a revenue-sharing telecoms concession into an excise tax payment that imposed a very considerable loss to the Thai exchequer.[26] Even if the substance of these findings is muddied

---

[25] 'Indonesian media tycoon juggles business and politics', *Nikkei Asia*, 3 March 2016.
[26] Thai News Service, 7 March 2018, and *Christian Science Monitor*, 26 February 2010.

by their highly political nature, it is very clear that use of his family network was a cornerstone of Shinawatra's operations. Other than involvement in his businesses, his brother was an elected member of the National Assembly while his younger sister – Yingluck – was appointed as prime minister in 2011 as Thaksin's apparent place-holder, before being ousted herself in 2014.

In addition, other close connections were able to benefit from public contracts. A notable case was that of Vichai Srivaddhanaprabha – best known outside the country for his ownership of the UK foot-ball team, Leicester City, at whose King Power stadium he was killed in a helicopter crash in 2018. Around that time, Forbes magazine reported him to be worth US$4.9 billion – the fifth richest man in Thailand. His King Power business was originally an agent for luxury brands and a duty-free outlet in Hong Kong. But its fortunes dramatically improved when in 2006 it secured exclu-sive rights to run the duty-free shops in Bangkok's airport through a tendering process that was widely criticised at the time.[27] Despite Shinawatra's ousting in 2006, King Power was subsequently able to build strong ties to other parts of the political spectrum and to maintain its concessions and sources of revenue.

The military junta that then assumed power has not itself been immune from multiple charges of impropriety. In 2019 a $7 billion high speed rail contract was awarded to a consortium led by the Charoen Pokphand (CP) Group, Thailand's second largest and a highly diversified business group, generating around $17 billion in revenue in 2017. This was despite the fact that the group had little, if any, experi-ence with railways. It did, however, have very close links to Chinese companies and the Chinese government[28]. Since the late 1970s, the CP Group invested massively in China becoming the largest foreign lessee of land in China with more than 50 per cent of the group's agribusiness profits coming from China. It was also the first foreign investor in the fledgling special economic zone of Shenzhen.

---

[27] *The Nation*, Bangkok, 14 July 2017.   [28] *Digital Journal*, 16 May 2019.

The CP Group is controlled by the country's richest family – Chearavanont. The patriarch – Dhanin – was Thailand's only billionaire in 2000 and 2008 when his net wealth was estimated at $1.2–3 billion. As we discuss in Chapter 4, by 2020 this had ballooned to $13.5 billion, whilst four other members of the extended family had a further combined wealth of nearly $12 billion.[29] Aside from involvement in their highly diversified portfolio of businesses, members of the family have been members of parliament and/or acted as advisers to ministers as well as maintaining close connections with both politicians and the military. Although formerly part of Shinawatra's inner circle, a sign of the patriarch's adaptability was that he was even being touted in 2019 as possible prime minister of a coalition government.

Networks linking business and politicians are not necessarily only internal, national phenomena. There can be a cross-border dimension. One such example brings together connections in both India and Malaysia. The Indian telecom company – Aircel – was sold in late 2005 to a group – Maxis – controlled by a Malaysian businessman – Ananda Krishnan – a man with strong connections to the political establishment. It was alleged that the transaction had been prepared at the Indian end by pressure that the Central Government Minister of Telecommunications – Dayanidhi Maran – had exerted on Aircel's owner, C Sivasankaran.[30] That pressure manifested itself as denial of necessary permits or licenses. Yet once the sale of 74 per cent of Aircel for ₹40 billion was consummated, those permits materialised in short order for the new owners. In addition, Maran's brother – Kalanithi Maran, the owner of a large Indian media company, Sun Direct TV – received an investment of around ₹55 billion, then equivalent to $1.25 billion. An additional feature was that the then Indian Finance Minister was found to have held up approval of the deal until his son received a 5 per cent share.

---

[29] Forbes World's Billionaires List (2020).
[30] *Economic Times*, 8 January and 1 September 2015.

The case has, in the manner of legal procedures in India, been long strung out. But what it illustrates very crisply is the way in which politically connected business transactions can play out. First, it highlights the critical influence that politicians and regulatory agencies can play due to the need to gather permits or licenses to operate or expand. The discretion that this affords politicians when regulatory authority is vested in them or where the regulatory agency is effectively captive can be very substantial. Second, it shows how businesses may prosper directly from having politicians in their fold, preferably within their immediate family. The case of the current generation of the Maran family in the southern Indian state of Tamil Nadu is instructive. They form part of a wider political dynasty. Their father was a minister in the central government; a close relation, Karunanidhi, was formerly chief minister of Tamil Nadu while his son – the ominously named M.K. Stalin – has also been a major figure in Tamil Nadu's politics for many years. One of the brothers – Kalanithi – migrated from politics to business, building a substantial media empire – the Sun Group – along with personal wealth estimated in 2020 at over $1.6 billion while also, as this case illustrates, leveraging his political relations and connections for his business' benefit. Third, the Maxis–Aircel case suggests that politically connected businesspeople may be attracted to connecting with others of the same ilk. The Malaysian dimension in this instance, Ananda Krishnan, has had long-standing connections to Mahathir and, at one stage, to Malaysia's Petronas. He has also been on record loaning the infamous 1MDB fund a significant amount. It has been alleged that the loan was part of some sort of bargain with the then prime minister – Najib Razak – to ensure that he would rebuff attempts by the Indian authorities to extradite Krishnan to face trial in the Maxis–Aircel case.[31]

---

[31] www.gfilesindia.com/product/september-2016 and *The Straits Times*, 14 February 2015.

3.5.3   Straddling Political Regimes: The Military

Military networks exist irrespective of political regime, although many of those networks have been built and consolidated in periods of autocratic rule. As such, the military often is able to maintain powerful networks and assets in democracies. For example, in Pakistan, even with an elected government, the military currently controls over a hundred commercial entities through five military welfare trusts. The largest of those trusts is the Fauji Foundation whose annual revenues have exceeded $1.5 billion in recent years. Its many business ventures include operating a security force, an oil terminal and a phosphate joint venture with the Moroccan government. Another – the Army Welfare Trust – runs one of the largest lenders in Pakistan – Askari Commercial Bank – as well as an airline and a travel agency. The National Logistics Cell is Pakistan's largest corporation and the country's largest shipper and freight transporter. It is also involved in road and bridge building as well as storing a significant part of the country's wheat reserves. The most detailed study of these trusts estimated that the net worth of the companies they control at more than $15 billion. Further, the army alone owns around 12 per cent of Pakistan's land even as wider military interests control at least a third of heavy manufacturing. Aside from the size of their interests, most military controlled companies operate outside of public scrutiny; barely 10 per cent of the businesses controlled by the five trusts are listed.

In Bangladesh, a similar – but smaller-scale – model has been pursued with the army expanding its business interests through two welfare foundations, the Bangladesh Welfare Trust (AWT) and Sena Kalyan Sangstha (SKS). The latter alone is estimated to hold assets valued at over $700 million with involvement in a wide range of activities including cement, food, electronics, flour and textile mills, filling stations, trading, insurance and real estate. AWT assets include several luxury hotels, golf clubs, a shopping complex, a commercial bank and – somewhat ironically – the firm that manufactures the voting machines used in national elections for which there was no tendering for the contract!

The military's involvement in the economy is by no means limited to South Asia. In Thailand, although the explicit involvement of the military has been reduced, the reality is that the institution still has a powerful role in the economy. For example, the TMB Bank – previously known as the Thai Military Bank – was originally established in the 1960s for the benefit of military families. In due course, it became a conduit for providing loans to dummy companies owned by members of the military. In 2005, the bank became a civilian institution with the Ministry of Finance as its largest stakeholder. Even so, the head of the army continues to serve on the board of directors and the Thai military is still a major shareholder in the bank.

Following the military coup of 2014, the military's involvement in the economy has again risen. The Royal Thai Army is presently involved in at least fifteen types of business ventures, including two television stations and over one hundred radio stations, as well as petrol stations and golf courses. It has a highly diverse portfolio that also includes indirect ownership of horse-racing tracks, shooting ranges, restaurants, road construction ventures, convenience stores, flea markets and boxing stadiums. Not surprisingly, there is a lack of transparency and audits of the military budget. Moreover, the military has cultivated close links with business setting up, for example, a large social enterprise in collaboration with leading businesspeople and companies. Furthermore, it has been estimated that out of fifty-six state-owned enterprises, forty-two have directors that come from the military. Serving officers are also quite commonly directors of private companies.

In Indonesia, Suharto's long rule involved pervasive and direct military involvement in the economy. Explicit policies to limit this involvement have subsequently been put in place, although this has not stopped military officials holding indirect stakes through trusts and foundations. And, as we have already noted, some of those with military pasts and ties to Suharto continue to operate as politicians.

In China, also, major companies are owned or controlled by the Chinese military. One recent list compiled by the US Treasury includes Huawei, Hangzhou Hikvision Digital Technology, China Railway Construction Corporation, Sinochem as well as other prominent state-owned enterprises. Vietnam's largest telecommunications service provider, Viettel, is owned by the Vietnamese military with the parent organisation being the Ministry of Defence. This has enabled Viettel to get access to land, infrastructure, technology and labour. Another example is the Vietnamese Navy's control of the Saigon Newport Corporation which has a pivotal role in external trade. It is the largest container terminal operator having a 90 per cent market share in Ho Chi Minh City and a 60 per cent share at the level of the country.

In short, the military's tentacles extend through many Asian countries and involve substantial direct and indirect ownership or control over companies and profit-making activities. Those tentacles' reach may generally be greater in autocracies – or countries which have had a prolonged autocratic past – but once established they tend to be persistent and highly resilient. In fact, the phenomenon of persistence is something that is more widely the case and to which we now turn.

3.6 LONGEVITY BUT ALSO TURNOVER IN THE RANKS

A striking feature of the wider Asian connections world is the way in which connections tend to persist, even when patrons and clients change or lose power. As we will see in Chapter 5, at its most accentuated, longevity and its bedfellow – entrenchment – can hold back the emergence of new companies and impede innovation and dynamism. Possibly the most pronounced example is Pakistan, where several deeply entrenched networks of interests – some overlapping – have continued to dominate since the country's creation in 1947. The two most notable forces have been the military – as we have just noted – and large landowners, particularly in the populous states of Punjab and Sindh. A significant part of the funding for the various political parties has continued to come from landowners – particularly those involved in the sugar industry and trade. Most of the eighty-odd sugar mills in the country are owned by political families – often dynasties – that have proven very effective in mobilizing large vote banks in rural areas.[32] Those families also commonly elect members to Parliament in their own right. Aside from funding political parties, the lobby has benefitted for decades from various subsidies, rebates and preferential tax treatment. In recent years, most – nearly 60 per cent – of those subsidies, including freight subsidies on sugar exports, have been received by groups of companies controlled by only three families.[33] In 2020, the CEO of one – RYK Group – had his brothers working as ministers in both the central and provincial governments. The owner of another major sugar company – JDW Group – has been instrumental in cobbling together votes and parliamentarians for the political party that won the elections in 2018, thereby allowing them to form a government.

In addition, the family business of a former prime minister – Nawaz Sharif – which owns six sugar mills received nearly 6 per cent of the total subsidies. Sharif's family has an extensive involvement in politics – including, inter alia, his brother, daughter, son-in-law and nephew – and their commercial and industrial interests encompass not just sugar but also transport, agriculture and steel mills. Disbarred from office and subsequently sentenced to jail on account of corruption, Sharif exemplifies the way in which Pakistani political and business networks have remained entwined. Among the many consequences of the entrenched power of landowners, a small – and sometimes overlapping (as in the case of the Nawaz family) – coterie of industrialists and the powerful military, Pakistan has failed to spawn sufficient new companies, let alone innovative ones.

Entrenchment does not necessarily equate to stasis. In much of Asia, despite the fact that connections are often persistent, it is not

---

[32] www.ft.com/content/0dcae657-3964-4951-8e63-bbaa86e1d893.
[33] Pakistan Enquiry Committee Report on Sugar Prices (2020).

generally the case that the world of connections precludes significant turnover or entry by new players. Moreover, loss of connections – through, for example, a change in government – is not necessarily tantamount to business suicide, as the persistence of so many business groups charted in Chapter 4 testifies, despite multiple changes in government. Conversely, connections can also ensure that some form of parachute is provided for politicians' losing their seats or positions. This may arise through appointments to company boards – a practice common across many countries – but also by less visible processes. In sum, what is so striking is the way in which the mutation of connections can occur without necessarily disturbing the underlying equilibrium and functioning of the system.

The Philippines presents one of the most accentuated cases in which the links between politicians, political parties and businesses have been powerfully present for many decades. During the Marcos regime in the 1970s–80s, for example, the business empire of Marcos' close ally, Eduardo Cojuangco, was thought to account for 25 per cent of the country's national income. But he has not been alone. The longevity of these connections is reflected in their dynastic nature. In fact, locals often speak of ten families thought to control the economy and the country's politicians. These include the group of companies controlled by the Ayala family, the origins of whose fortune date back to the 1830s. As with the other major business groups in the country, this group is hugely diversified, currently holding stakes in telecommunications, real estate, education, financial services, utilities, transport and infrastructure, as well as healthcare. The Alliance Global Group, controlled by the Tan family, has holdings and interests in media and entertainment, transport and infrastructure, food and beverages as well as real estate. The San Miguel and SM Investment groups – also family controlled over several generations – have interests in many sectors including extractive (mining) industries.[34] Table 3.3 indicates the very diversified nature of

---

[34] asia.nikkei.com/Spotlight/The-Big-Story/Crony-capital-How-Duterte-embraced-the-oligarchs.

Table 3.3 Areas of operation for main business groups in the Philippines[1]

Sector	Ayala	Alliance Global	LT Group	Metro Pacific	JG Summit	San Miguel	SM Investments	Udenna
Banking	*		*		*	*	*	
Education	*					*		
Food & Beverage		*	*		*	*		
Healthcare	*			*				
Media		*					*	*
Mining		*					*	*
Power & Utilities	*	*	*	*	*	*		
Real Estate	*				*	*	*	*
Retail	*						*	
Telecoms		*		*	*	*		*
Transport & Infrastructure	*			*				*

[1] Nikkei Asia (2019b).

the major Filipino business groups' interests. These diversified interests have been built up over time through a process of accretion and the seizing of opportunities. Many of those opportunities have come from connections to power and ruling parties or governments. Yet, aside from being around for decades, these groups have sustained themselves – and often grown – even when the political regime has changed and those in power have turned hostile. Some of the business groups' fortunes have varied – occasionally quite theatrically – depending on the government of the day. For example, the current president (Duterte) recently under-mined one well-connected and powerful incumbent – Roberto Ongpin – and with him, his group of companies. Shortly after, his attention turned to the Ayala and First Pacific groups, owned, respectively, by the Ayala and Pangilinan families, over a dispute about water utilities, as well as revoking the license of the largest broadcaster – ABS-CBN – itself owned by the powerful Lopez family.[35] In all instances, it was projected politic-ally as an assault on the power of the local oligarchs and their allegedly abusive behaviour. Faced with this hostility and a sharp fall in its equity value, the Ayala Group sold control of its water business to another businessman in better standing with the president. Whether fortuitous or not, both groups' donations to public health in the pan-demic of 2020 induced an apparent change of attitude to them by the president.

As we shall shortly see with the case of the Salim Group in Indonesia, a loss of political favour does not necessarily presage doom – and almost never does once a business group has become entrenched. Even so, being out of favour does commonly lead an adversely affected business or business group to shift resources out of the country, mostly into neighbouring economies, while waiting for a change in the political climate at home.[36] As such, the

---

[35] www.ft.com/content/c295d548-51b0-11ea-90ad-25e377c0ee1f.

[36] If the political shift is sustained, companies have then the option of fully relocating; Witt and Lewin (2007).

pressures placed by governments – often when they initially come to power – tend to result in – at best – a partial changing of the guard. Rather than purging what is often described as an oligarchic system, space has simply been created for new incumbents that are more closely aligned with – and dependent on – the current administration.

The most visible recent example in the Philippines has been the Udenna group founded and run by Dennis Uy who hails from the same city as the president. Again, with diversified interests in petroleum, shipping and logistics, the group has also rapidly expanded into, among other things, telecommunications, real estate, retail and media and entertainment. Its dramatic growth appears to have been fuelled by a massive accumulation of debt. This borrowing – mostly from domestic banks – has itself been facilitated by the political connections of the group's founder. The spectacular rise has seen Dennis Uy become one of the richest individuals in the Philippines with a net wealth of $660 million by 2019.[37]

West of the Philippines across the South China Sea, Malaysia has acquired a prolonged – and well documented – history of close links between politicians and business. The longevity of some of the key political figures themselves has also accentuated the impact of such connections. In essence, several broad types of connection have been present. One strand has related to the explicit policy of promoting Malay-owned businesses (Bumiputras) that was pursued for several decades from the mid-1970s onwards. This was an explicit attempt to create local rivals to businesses owned by people of Chinese or other origin. Probably the most well-known example of this policy was the ill-fated attempt to create a domestic car industry under the Proton brand. The principal Malay shareholder – Syed Mokhtar Albukhary – was closely connected to the then Prime Minister, Mahathir Mohamed. Because of this link, he became a massive beneficiary of public support,

---

[37] www.forbes.com/profile/dennis-uy/.

including a later – and very substantial – bailout in 2016 to address the company's persistent losses.[38] Another form of connection has mainly involved links between individual businesspeople and a fairly small group of dominant politicians in the then-ruling party. Some of those links dated back to their schooling together. A study of political power and its concentration found that the key players numbered no more than four to five politicians.[39]

A third strand relates to businesspeople of Chinese origin. In fact, this is a feature through much of Southeast Asia. In Malaysia, Robert Kuok is the most notable example. Born in the state of Johor, Kuok's father had already put down strong business roots in that state having received a license from the local Sultan. Kuok took over and expanded his father's rice, sugar and wheat flour trade by capitalizing on his work with the local elites in government. Kuok had been at school with both the future Malaysian prime minister and Lee Kuan Yew who later became Singapore's long-standing prime minister. These connections ultimately proved very helpful. Kuok's first large ventures were in the 1960s with Malaysia's first sugar-refining company, Malayan Sugar Manufacturing. With some preference from government, it soon built market dominance.[40] Kuok also established the country's first sugar cane plantation – Perlis Plantation – on state-owned land and this was followed in the 1970s by more sugar plantations in Indonesia. The scale of operations was such that at times he controlled up to 10 per cent of the world sugar market. At all times, close proximity to government was maintained and not just at home. His close association with a fellow Chinese, the founder of the Salim Group, helped establish close links to the Suharto government in

[38] In 2017, China's Geely acquired a 49.9 per cent stake in Proton and provided significant new investment in the company. www.reuters.com/article/proton-strategy/malaysias-proton-expects-first-profit-since-deal-with-geely-ceo-idUKL4N28N1G5.

[39] Gomez and Jomo (1998).

[40] www.straitstimes.com/asia/se-asia/malaysian-govt-gave-kuok-key-to-become-sugar-king/.

Indonesia. The pattern was repeated in Myanmar.[41] But probably the most strategic decision was to invest early in China – something also done by Thailand's CP Group. This allowed him to build many close links to the leadership of the CCP. In the future, aside from doing business in China, it also permitted Kuok to act as an intermediary between China, Malaysia and other countries, as well as the Chinese diaspora.

In Kuok's case, overseas expansion was also accelerated as a means for mitigating political risk when the Bumiputra policy was adopted. At the same time, the business was diversified further with moves into real estate along with beverages, industrial manufacturing, shipping, investment and insurance, media and hotels. By the end of the 1980s, there were around 250 companies controlled by Kuok, his relatives, trusts and obscure shell companies in a variety of jurisdictions on- and off-shore. Since then, as its holdings have expanded substantially, its finances have become even more arcane. A measure of Kuok's resilience is that when Mahathir – the architect of the Bumiputra policy – returned to power in 2018 after the downfall of Razak due in large measure to the 1MDB scandal, Kuok was appointed to sit on a special council of advisers to the government whilst also continuing to act as a conduit between the government and China.[42]

The ties that have bound a relatively small group of politicians to a set of businesses and business groups in Malaysia have spanned a variety of mechanisms, ranging from import substitution – and hence preference given to local companies – to participation in industrial policy projects, as well as more generally benefitting from public contracts and funding, often from state-owned banks. Even some macroeconomic policies that were adopted appear to have had a strikingly beneficial effect on politically connected companies.

---

[41] Cheong et al. (2015).
[42] www.scmp.com/week-asia/politics/article/2145849/hong-kong-billionaire-robert-kuok-advise-mahathir-during-his.

Despite a sharp, initial loss in the value of companies seen as closely connected to politicians at the start of the Asian crisis in 1997, this was handsomely reversed in the aftermath of capital controls being introduced in September 1998. At that point, those connected companies or groups experienced a $5 billion increase in their market value, of which around a third has been attributed to their political connections. The value of those connections might have been as high as 17 per cent of their total market value.[43]

How connections work and their longevity in Malaysia is, of course, far from singular. In neighbouring Indonesia, before the ousting of its autocrat, Suharto, in 1998, the country had often been feted as a shining example of development. Yet, it was clear that there were very close links between Suharto, his family, coterie and a number of favoured business groups. Consider the case of the Salim Group. Founded as a small-scale trading firm in 1938 by a Chinese merchant, Liem Sioe Liong, its spectacular, later growth can be traced to the founder's relationship with President Suharto.[44] In the course of the latter's long rule and with a deft use of business partners, including some of Suharto's relations, the business prospered hugely as it benefitted from government contracts and other preferential deals, diversifying into many activities and sectors, including in other countries in the region. The group leveraged its connections to the autocrat to gain business but also to hedge its risks. In addition, family members became involved in shaping bits of public policy that could also help promote their interests. Some of that translated into a dominant position in markets where protection against competition was granted, such as the nascent and heavily protected local steel industry.

What effect did the fall of Suharto in 1988 have on the group? Initially, it was pursued by the new administration and fines and other penalties were imposed. It disposed of one of its main assets. Yet the group survived and avoided the outright dispossession or judicial

---

[43] Johnson and Mitton (2003).   [44] Dieleman and Sachs (2008).

penalisation for corruption that some had demanded. To adapt to the new circumstances, the group increased its investments in other countries in the region and withdrew from activities and contracts that were strictly dependent on political links or patronage. Although its influence and ability to exploit its connections waned after Suharto's fall, it was by no means forced to the margins or into the shadows. This was despite the emergence or growth of competitors with 'better' connections to the new governments. Indeed, this resilience is reflected in the fact that the Salim Group through its holding company – First Pacific Company – is still one of the largest business groups in Indonesia. One of its companies – Indofood Sukses Makmur – is presently the sixth largest company while an affiliated company – Indofood CBP – is ranked twentieth.

An early partner of the group – Sudwikatmono, a cousin of Suharto – has also maintained his position, with his Indika Energy currently ranked as the nineteenth largest company. The Sinar Mas group – another diversified entity with its main interests in palm oil and paper manufacturing as well as real estate, energy, financial services, telecoms and mining and annual revenues of $30 billion along with 380,000 employees[45] – also dates its rise back to the Suharto period when several of its business ventures were in collaboration with Salim. That association was critical in securing approvals and licenses for its projects along with funding from state-owned banks.[46] Currently, the largest – by far – company in Indonesia – PT Astra International – whose majority shareholding is held by Hong Kong-based Jardine Matheson Holdings – also had very strong links to the Suharto regime. The simple truth is that many of the businesses whose political connections provided them with huge opportunities in an autocratic period of Indonesia's history have subsequently continued to prosper and remain major planks in the Indonesian business world to this day.

[45] Nikkei Asia (2019c).   [46] Chen and Redding (2017).

These narratives underline how longevity – persistence – has been rife. Has this been attributable to the ability to find new political connections and patrons? The answer is not necessarily. It does not appear that a business or business group has to acquire a political patron of equivalent heft to their original patron, although doing so can be helpful. Rather, initial advantages can be entrenched if there is a continuing ability to influence the regulatory and competitive environment. By such means, connected companies continue to translate what might have been transitory preference into a form of permanent advantage. This path dependency can, of course, pave the route to incumbency. Yet, even as the connections world sustains itself, including in the face of new networks and entrants, the broader policy environment is rarely frozen in aspic. This means that although these networks will always try to condition or influence how such changes are designed or implemented, they must also have the flexibility to adapt. The capacity for adaptation and being fleet of foot is essential. What has unfurled in India illustrates this very well.

3.7 SELECTIVE ADAPTATION: FROM LICENSE RAJ TO LIBERALISATION IN INDIA

3.7.1 Land of Licenses

Since Indian Independence in 1947 there have been a series of different policy phases.[47] Initially, many of the commercial and industrial supporters of the ruling Congress Party were effectively granted rights to particular sectors or activities. One of the most visible examples was the Birla group's fifty-year near-monopoly of the indigenous automobile market through its increasingly antiquated Ambassador car. But this was by no means an outlier. Most major areas of economic activity were allocated in some manner to preferred family-controlled

---

[47] Panagariya (2008).

business groups. Other activities – such as aviation, the banking system and parts of manufacturing – were simply nationalised given successive Congress governments, promotion of a greater role of the state in the economy.

The period – spanning from shortly after Independence up to the 1990s – is commonly known as the License Raj. Activities, or indeed whole sectors, were allocated to specific companies, chosen mostly on the basis of their past, or intended allegiance, to the dominant political party. This allegiance often included making significant contributions to the Congress Party. One Indian prime minister of that time was upbraided in the press for receiving suitcases of cash from so-called industrialists. The response was simply that these were donations to the party. Consequently, much of the formal economy – that part of the economy which was taxed and regulated – was either directly in the hands of the state or in those of private, family-based business groups. These business groups – as in the other countries and examples that we have used – were generally very diversified. The dominant business strategy was to acquire or inveigle permits or concessions for activities granted through political patronage. Subsequently, these groups invested large amounts of capital and time in maintaining those privileges including, of course, narrowing or eliminating the scope for competition, whether from other local players or from imports. In short, the assembly of diversified portfolios mainly reflected various groups' ability to acquire and exploit licenses or concessions from government.

This strategy was radically undermined and in large measure dismantled by a programme of economic liberalisation, including lowered trade barriers and an opening to greater FDI, that was undertaken in the 1990s. These reforms made the search for, and monetisation, of licenses and permits less relevant and ultimately less attractive. With those changes also came a shift in incentives for large businesses to continue as diversified entities. Although some of the reasons – such as diversification of political risk – still existed, in general the assembly of heterogeneous portfolios gleaned from allocations made by supportive politicians was far less of an option. Moreover, even when connections could help secure assets, the chances were that they would be subject to far greater competition than had existed throughout the License Raj period.

These shifts are reflected in the composition of the 2020 list of Indian billionaires and the sources of their wealth, which we present in Chapter 4. Presently, a relatively small (less than 15 per cent) share of the total number of billionaires (amounting to over 100) control diversified businesses. Most of those that still do are families or individuals that assembled their businesses, or saw them grow, under the License Raj. Further, they mostly lag behind some of the newer and more dynamic businesses whose owners' strategies have largely been to concentrate on a relatively small core of activities.

3.7.2 *Modernising Connections*

Has the demise of the License Raj affected how politicians and businesses interact and how their respective networks are organised? The answers are quite nuanced. Clearly, the links that bound the then-ruling party to specific, privileged groups have been undercut. But the feature that keeps politicians and businesses in lockstep is that for many types of activities, government or its related agencies hold important regulatory authority and discretion. This may comprise granting access to land or mining rights through concessions, but also the allocation of spectrum for telecoms operators or aviation rules and permits, as well as the weight of public contracting for infrastructure projects, whether at provincial or federal levels. These overlapping spaces mean that for many businesses in India it is almost impossible to function – let alone succeed – without having political connections.[48]

---

[48] Crabtree (2018) discusses how politics and businesses have come together in recent decades.

Take the case of real estate where the principal issue is access to land and building permits. DLF Limited is one of the leading commercial property developers in India and a main force behind the extraordinary rise of Gurgaon; a new city neighbouring the capital, New Delhi. Forty years ago, it was a largely agricultural zone; by 2011 it had a population of nearly 900,000 which has now risen to 2.8 million. DLF's CEO – Kushal Pal Singh – built very strong ties to the former Prime Minister – Rajiv Gandhi – and his family, including subsequent business ventures with the latter's son-in-law. The company was able to acquire housing and commercial property development permits as well as an initial 3500 acres.[49] Over time, DLF's land bank has swelled to over 10,000 acres. Despite being handed out a massive fine by the Competition Commission for unfair pricing of goods and services in 2011,[50] irregularities in its 2007 IPO that had been identified had not been acted on. It was only when the government changed in 2014 and the BJP took power that these complaints were pursued. DLF's top executives, including its CEO, were barred from participating in the securities market for three years. This led to a very sharp fall (in excess of $1.3 billion) in the company's equity: a clear indication of the value of political connections.

While DLF's story illustrates the perils of being seen to be too closely associated with one political party – and in this case political dynasty – the company has invested significantly in trying to build ties to the governments, federal and provincial, with which it had hitherto been unconnected. Although DLF's share price fell in the wake of the Congress Party's loss in the 2014 elections, by the first quarter of 2020 it was higher than in the three years preceding the 2014 election. In addition, KP Singh's fortune is currently estimated to be worth around $3.7 billion making him the nineteenth richest

[49] in.reuters.com/article/india-dlf-kpsingh/banned-property-tycoon-kp-singh-who-built-indias-first-smart-city-idINKCN0I60WL20141017.
[50] in.reuters.com/article/idINIndia-58809520110816.

man in the country. This suggests that a loss of a central political connection and associated patronage need not necessarily be permanently disastrous.

3.7.3 Reaping the Fruits of Modernised Connections

Another company that has been widely perceived as building its business largely through its very close connections to government, in this case the current BJP government, is the Adani Group. These connections date back to the 1990s when Gujarat's then-Chief Minister, Narendra Modi, explicitly pursued policies that were favourable to business. Adani – along with others – benefitted from tax and other fiscal benefits, as well as access to land at low prices. Subsequently, the group has been able to secure large public contracts. These have included twenty-five successful bids for piped natural gas networks and fuel stations as well as winning all six contracts that were tendered for managing airports, despite having no prior experience in the field unlike competitors that were bidding.[51] Rather than consolidating its focus, the group has become increasingly diversified but, notably, in areas where government regulation and public contracting have been central features.[52] As a result, the group has grown very rapidly with combined revenues of over $13 billion and a market capitalisation of $22 billion.[53] On this rising tide, Adani's own wealth jumped by over 130 per cent in 2020–21 making him the country's second richest individual. Most of that increase has been driven by huge increases in the value of shares, mainly in his energy companies. But a significant part of that equity appears to be held by a small number of overseas funds with a very limited public float making them susceptible to volatile movements in value.

---

[51] Rajshekhar (2019b).
[52] Modi's Rockefeller? www.ft.com/content/474706d6-1243-4f1e-b365-891d4c5d528b.
[53] Rajshekhar (2019b).

The continuing aid that proximity to power confers in activities where government actively influences either, or both, the rules of the game, as well as contracting, is also well illustrated by the example of Mukesh Ambani's Reliance group. This is the most successful Indian family company of recent decades, and one that has forged very close ties to a succession of governments. Although the overwhelming bulk of Reliance's activities has remained in petrochemicals,[54] where its approach has been to be the dominant player across all ends of the crude oil chain – upstream and downstream – since the turn of the century, the group has branched out into telecoms and then into e-commerce.

Moving into telecoms has of course required close interaction with the Government of India. That is because governments everywhere must allocate band spectrum. Initially, Reliance acquired spectrum in 1999 from the Government of India with a further acquisition in 2012. The central part of the business strategy involved widening the use of the spectrum that had been acquired, a process that has of course been facilitated by the changes in the underlying technology and capabilities of mobile broadband. Reliance's initial acquisition in 1999 was of a small and obscure company – Infotel – that had successfully acquired spectrum, bidding over five thousand times its supposed net worth, only to be then acquired immediately by Reliance. This gave rise to accusations of collusive behaviour including in a report that was subsequently prepared by the Comptroller and Auditor General of India. Further, the spectrum that had been acquired was only permitted to offer internet service (based on an internet service provider permit) but it was soon translated into a full mobility or unified license which allowed Reliance to offer all mobile services but at a fraction of the cost that a full service license should have incurred. Some observers – including the Auditor General – have

[54] McDonald (2010) provides history of the Ambani family and Reliance.

suggested that the foregone revenue to the exchequer might have been as high as $4 billion.[55]

Using a strategy of rapidly building market share by offering highly competitive prices and options, Reliance has also proven adept at bargaining with the regulator on pricing. Lower access and inter-connection costs with other operators' networks have resulted while rules regarding predatory pricing have been changed in ways that could be seen to favour Reliance. By 2020 their customer base was approaching 400 million subscribers. As a consequence, the underlying valuation of Jio is already significantly higher than India's largest mobile operator – Airtel.

Reliance Jio's business strategy allows two complementary ways of improving profitability in the future. The first will be to translate market share into pricing power. The second will be to leverage and, ultimately, monetise the large volumes of data to which they have access. This will not be through a conventional telecoms model but by linking to a broader range of activities, such as food and fashion retailing, with very strong backward linkages in the form of infrastructure and media that themselves are data generating. Large recent investments by Facebook and Google (along with some major US private equity firms) have opened up ways of using WhatsApp, which has a major presence in India, to offer retail services as well as an Android-based smartphone operating system designed for the Indian market. As such, Jio aims to bring together the hitherto disparate parts of its operations and the company's links to government and its ability to project a national, even local, image, as opposed that of leading competitors like Amazon or Walmart, will confer important advantages.

Depending on the government for licenses, concessions or regulatory approvals is, by no means, uniformly required. For some sectors, such as pharmaceuticals, information technology and software and parts of finance, this dependency is largely absent. In the

---

[55] Anjumol et al. (2019).

case of India's most visible success story of recent decades – IT – companies face very limited regulation, mostly sell into foreign markets, have to have few dealings with infrastructure operators (such as ports) and have a high skilled, well-paid and largely non-unionised labour force. All these features necessarily lower the scope for political interference and rent seeking. In pharmaceuticals, some similar considerations are at work. Consequently, the main interactions of the leading pharmaceutical companies with government and politicians have been around intellectual property issues and access to foreign markets. In short, the waning of the License Raj has undoubtedly lowered the relevance of connections for some activities but the scope for such connections still remains important, as the examples we have used illustrate all too well.

3.8   THE RECIPROCITY GAME

As we mentioned in Chapter 1, the connections world is most definitely made up of two-way streets. The reciprocity game can be seen in many guises. Of course, the most obvious manifestations are contributions to political parties or politicians, even if the degree of legality, let alone transparency, of such contributions varies a great deal. But there are also other important and costly actions that connected companies tend to make. In Indonesia, the Salim Group had to periodically bail out, take stakes or do business with entities or persons that were relatives of Suharto. This was clearly understood as the quid pro quo for the benefits conferred by Suharto and his administration. When the Salim Group's steel company was failing in the early 1990s, it was the turn of the Government to bail it out. Mutual scratching of backs is a hallmark of many such connections. In the Philippines, the rapid rise of Dennis Uy's business group also displays some characteristic features of the connections world. The group has made substantial contributions to political campaigns. When awarded a new license to challenge the existing telecoms duopoly (both controlled by incumbent families), Uy's group helped deliver on a key campaign promise of the president that a new entrant would be permitted. Yet this license realistically offered few, if any, short-term financial rewards but was, rather, important to the group from a strategic perspective.[56]

Among other recent reciprocity plays have been the various responses of businesses and business groups to the COVID-19 pandemic. In Vietnam, the richest person – Pham Nhat Vuong – has used his Vingroup to produce ventilators under license from an Irish company while also allocating over $5 million to the acquisition of medical supplies as well as testing and research.[57] The group's retail arm – Vincom – allocated over $13 million to support tenants affected by the COVID-19 pandemic. In the Philippines, several of the main business groups have also contributed significant amounts to public health spending in the pandemic and, as we noted earlier, some have already reaped a clear political reward: a cessation of hostilities from the current government. In India, the well-connected JSW Group made a large donation to the Prime Minister's Emergency COVID fund.[58] In Thailand, the owner of the leading hospital network donated $1.25 million in 2017 to the government department in charge of prisons for use with ex-offenders. In similar vein, the largest Indian mobiles company – Airtel – has contributed in a major way to the government's attempts to improve sanitation in rural areas.[59] In fact, there is a long list of major companies and business groups making these sorts of donation and public investment across Asia.

Donations to politicians' preferred causes, or at times of national emergency, are a common, almost standard, feature of the connections world. Less visible, of course, is the way that businesses create jobs for ex-politicians and officials or, indeed, friends and relations of their political connections. The Adani Group, as well as other

[56] Nikkei Asia (2019b).
[57] www.bloomberg.com/news/features/2020-06-08/vietnam-s-richest-man-plans-ventilator-exports-for-covid-19-cases.
[58] www.thehindu.com/news/national/jsw-group-commits-100-crore-to-fight-covid-19/article31198615.ece.
[59] economictimes.indiatimes.com/news/politics-and-nation/bharti-airtel.

major Indian business groups, makes a habit of hiring former officials
with exposure to the regulatory and licensing needs of their companies. In fact, this approach is widespread across Asia with well-
oiled, revolving doors between government and business. Further, as
we shall see later in Chapter 6, the creation of jobs or, conversely, the
limitation of job losses in response to the interests of politicians is an
important strand of the wider reciprocity game. In sum, although the
motivations may be different, maintaining or creating connections
with government is the hallmark of reciprocity. It is a form of
acknowledgement of the benefits that such connections confer.

3.9   CONCLUSION

The connections world is exceptionally well embedded in Asia. The
main players – businesspeople, business groups, politicians and political parties – are present throughout, although the exact configuration
of the networks depends on how the wider political system and
associated institutions are arranged. The resulting differences in
how connections are applied and the channels that are used are clear
when contrasting India and China, for example.

Yet, whatever the local variation and colour, these webs of
connections bind together with common purpose. The leveraging of
connections for mutual benefit has often delivered large benefits that
have often persisted across changes of government, let alone political
regime. In today's Asian democracies, many large, sometimes dominant, businesses – including those belonging to the military – were
built on the connections established in earlier autocratic eras.
Perhaps most importantly, these webs of connections have created a
system of behaviour and rewards that has become increasingly
entrenched. That entrenchment is as true in some of the richest
Asian economies, such as South Korea, as it is in some of the poorer
ones, such as Pakistan.

What is also clear is that the connections world carries important implications for how businesses organise themselves and how

they seek to function. One common feature we have observed in this chapter has been the emergence and subsequent persistence of family-based business groups. As we shall see, this form of organisation is by no means common throughout the globe. Indeed, it is a relative rarity in the rich world. Yet in Asia, it is almost the norm. As to why that is the case, and the consequences of it being so, is the subject of the next chapter.

4 Networks, Connections and Business Organisation

We have provided a fresh perspective on Asia's undisputed economic successes of the past forty years by viewing it through the lenses of connections, both those between businesses and politicians and those between businesses themselves. Despite the great diversity of political system, economic structure, histories and geographies across Asia, our analysis starts from the view that there is one central commonality – the pervasiveness of close business and political connections – throughout the region. Moreover, while this connections world has been an important element in Asia's rapid economic renaissance, it is now also increasingly becoming a brake on potential future success.

One aim of this book has been to characterise the connections world, using new data sources to provide striking visualisations of the connections world. There are variations in how the networks look and behave in different countries, but what is common is the inescapable presence of close and deep ties within the business community, between them and the political classes, and of the complex web of reciprocity between them. Chapter 3 not only provides data to characterise these relationships but enriches our understanding of how this shadowy and opaque world functions through examples and cases, some egregious.

As we turn from the nature of the connections to the actors who are connected, one type of organisation is increasingly manifest. Business groups are a way of structuring businesses almost uniquely suited to the connections world. Their opaque ownership arrangements and their non-transparent accounting processes facilitate legal, semi-legal and outright dubious relationships, both with other firms and also with politicians and civil servants. Moreover, their structures ensure that the development process acts disproportionately to enrich the oligarchic dynasties that own them. Most importantly, these business groups are entrenched with huge monopoly power across the bulk of resource, capital and product markets of the region. Often working with their political connections to influence competition and trade policies, they exploit their market dominance to restrict competition and new firm entry, and to choke off innovation. While their scale and diversity may have made them important mechanisms to mobilise resources in the early phase of economic development, they have now become the vehicle whereby the connections world can act to limit the next stage of economic development.

Many of these issues are illustrated when we turn to Asia's performance in innovation, a crucial element in future economic growth. We consider how the pace and quality of both the development of new products and the implementation of new processes is affected by the connections world. The question is what happens when science, innovation and national champion policies to accelerate innovation collide with the vested interests of the connections world? Can Asian business groups go beyond entrenchment and rent-seeking? The answer is necessarily nuanced and differs somewhat by country, but the incentives to innovate are seriously compromised for business groups that are already in dominant market positions and the resulting market power allows them to erect barriers to entry by innovative new ventures. These barriers are made even higher to scale when supported by regulations and policies introduced by the political connections. Even when business groups do innovate, as when under pressure from international competition or when establishing themselves in domestic markets against rivals, they tend to revert quickly to type. Some innovative new ventures have emerged and have grown rapidly but this has mainly been in new industries and in large economies where the obstacles are weaker. But even then, the very success of these new ventures often quickly attracts the unwelcome attention of the connections world. The latter mostly aims to incorporate these upstarts into the system or subject them to control. Putting these strands together, there are powerful reasons for thinking that the power and influence of the connections world will be a serious brake on the ability to innovate in the future.

The Asian connections world that emerged in the previous chapter was seen to involve politicians and businesses coming together in ways that serve their mutual interests. In so doing, the network dimensions of these connections are prominent. These dimensions also extend from, and into, the very ways in which business functions. As such, Asia has developed some very particular forms of business organisation. It is with these particularities, their origins and consequences, that this chapter is concerned.

It only takes a side glance to pick up the very different organisational formats that separate businesses in Asia from their counterparts in the Western world. In North America and Europe, the bedrock of the business sector, the relatively few large firms that produce the bulk of output, is the public company for which capital markets and managerial control are meant to guarantee the primacy of shareholder interests and where there is, mostly, separation between ownership and control. This structure enables these corporations to raise large amounts of capital for investment aimed at raising productivity, without being constrained by their own cash flow or by the wealth of their owners. Listed companies are consequently held by a diverse group of shareholders – and run by a professional class of business managers. This separation of ownership and control and the associated need for owners to monitor managerial performance has also meant that companies tend to concentrate on a single industry or market, even if strategic rivalries can sometimes lead them also to move up- or downstream in their supply chains.[1] Thus, most successful modern corporations – think of Apple[2] – stay focused on one industry; indeed,

---

[1] See Wernerfelt and Montgomery (1988); Bowen and Wiersema (2005).

[2] Apple was founded in 1976 and went public in 1980, at which point ownership and control were largely separated. It has largely remained within its original remit of

capital markets impose a significant discount on the valuation of conglomerates.

The situation in Asia is fundamentally different. There we find that most firms, both small and large, are family businesses. Moreover, a high proportion of output is produced in a small number of business groups. These are loosely affiliated groups of firms, some publicly listed and some private. Business groups are best described as alliances of firms that are bound together in both formal and informal ways, often hard to discern, including, but not exclusively, ownership and usually through family ties.[3] They are distinguished by their family or dynastic ownership which is often opaque along with impenetrable organisational structures and governance arrangements. These business groups rub shoulders with numerous state-owned firms and together these dominate the Asian economic landscape.

In short, as we saw in the previous chapter, business groups are a hallmark of the region, and they are often deeply entrenched, both in their markets and within their political networks, as well as being very large relative to the economies in which they are based. Some of them have survived in this form for generations, even centuries. Business groups are also very diverse in terms of sector giving them control of considerable capital and labour resources along with clout across much of the economy. There is a superficial similarity with diversified Western corporations such as 3M or General Electric but business groups in the sense we use the term were actually made illegal in the United States by President Roosevelt after the Great Depression. As to why this form of business organisation has become so prevalent in Asia can be traced to multiple reasons, including lower levels of economic development and weak

---

personal electronics with some expansion up- and downstream. Revenues in 2019 were $174 billion with 137,000 employees.

[3] Granovetter (1994).

market-supporting institutions. But whatever the reasons, non-transparent ownership and governance structures have arisen and persisted, sometimes as an instrument of government policy. Above all, business groups are organisations that are particularly well-suited to succeed in the connections world that we identified in the previous chapter. Further, this adaptation and proliferation across Asia has had some enormously powerful consequences, some of which threaten to be very long lasting indeed.

We start this chapter by characterising these emblematic Asian business organisations including exploring the reasons for their rise and persistence.[4] In addition, the state has played a more interventionist role in the economy in Asia, including through widespread ownership of firms. Turning then to the question of how well business groups and state-owned enterprises perform, there is little doubt about the disadvantages of state ownership, but the matter of business groups is more complex. For example, their affiliates may perform better than non-affiliates in some respects, but the non-transparent transfer of funds – which we referred to in Chapter 3 as tunnelling – is also common. Moreover, our main concern is with business group entrenchment into the body politic and the impact on market efficiency, competition and later – in Chapter 5 – on innovation. Indeed, we argue that the influence of business groups is often malign, operating through market power, not only within individual markets but also at an aggregate level across markets. To capture this, new measures of the overall market power of business groups are introduced. We also discuss the implications of this concentration for inequalities in wealth which, as we have seen, have ballooned with economic growth. We conclude that the entrenchment of economic power built on connections and around business groups represents a major barrier to broad-based economic development in Asia.

---

[4] There is a voluminous literature, and initial reading might include Granovetter (1994); Claessens et al. (2000); Bertrand et al. (2002); Carney and Gedajlovic (2002); Khanna and Yafeh (2007); Carney (2008); Colpan et al. (2010); Carney et al. (2018).

4.2   BUSINESS ORGANISATIONS IN ASIA

In Asia, not only is the role of the listed company less prominent but there is a central role for family – often dynastic – ownership alongside control in both privately-held and listed companies. As a result, minority, or even majority, external shareholdings are typically combined with concentrated, strategic ownership by families or founder-managers. The institutional vehicles that have been adopted to maintain this control are also often highly complex, with convoluted and opaque cross-holdings and pyramidical ownership structures. It is rare that business groups are organised in a simple way around majority ownership by a family across a number of different firms. Rather, the controlling family often owns a majority stake in several other companies, which they use as vehicles, along with the family holding company, to own jointly majority stakes in other companies, and vice versa.

Let us start with the basic building block, family ownership. To get a sense of how widespread the phenomenon is, in India, Pakistan and Malaysia around half of all publicly-listed companies are family owned while in Indonesia, Thailand and Taiwan that share ranges between 58 per cent and 68 per cent. In Hong Kong and South Korea this rises to 81 per cent and 87 per cent, respectively. Even in China, nearly a third of listed companies are family-owned. To place this in a wider context, the equivalent share for the United Kingdom is under 25 per cent.[5]

But the differences with the Western world do not stop there. Most major Asian businesses – both family but also state-owned companies – are also organised in business groups; a stark contrast with the more common stand-alone model that exists in the advanced economies. These business groups come in many shapes and forms but are usually alliances of firms typically linked by ownership held directly and indirectly by one or a few families. Different terms are

---

[5] Anderson and Reeb (2003); Poutziouris et al. (2015).

used to connote business groups across countries, for example, *chae-bols* in South Korea, *hongs* in Hong Kong, business houses in India and *guanxiqiye* in Taiwan.

Despite different appellations, most business groups have many common features including much-diversified portfolios of activities commonly operating across many sectors. The example of the main Filipino business groups in Chapter 3 shows how diversity is a hallmark of all. There are many other examples. The second largest firm in Thailand, Charoen Pokphand (CP) – which we also encountered first in Chapter 3 – was founded in 1921, originally as an animal feed food company, but over time it has been transformed into a group of linked companies operating in telecommunications, petrochemicals, vehicles, property and finance. Similarly, Malaysia's Hong Leong Financial Group, founded as a trading company in 1963, although centred around its banking and financial interests, also has ownership of a highly diverse portfolio of companies in manufacturing, distribution, property and infrastructure development. This preference for diversification reflects not just the ways in which opportunities for expansion have arisen but also a set of deliberate responses to political connections and the way in which these groups interact with politicians and public agencies, not just to assemble their portfolios but also maintain and extend their interests.

As these examples indicate, business groups are not just a recent phenomenon: some of the leading Asian business groups have been around for a long time. The business group organised around the Ayala family in the Philippines dates back to the nineteenth century. One of Hong Kong's largest business groups – Jardine Matheson controlled by the Keswick family – has its origins in the same century, building its fortunes on trading Indian-produced opium into China. Its interests have subsequently spread around the region; as we saw, the group is now the majority owner of the huge diversified Indonesian business group, Astra. These features of long-lasting family control, along with diversified interests, is perhaps most clearly exemplified in India's Tata group, also founded in the second half of the nineteenth century. Around two-thirds of Tata Sons are still owned by family members or philanthropic trusts set up by members of the Tata family. The two biggest shareholders, both family members, own around 50 per cent of the shares and the Board has almost always been chaired by a member of the Tata or Mistry family since its foundation. This complex structure of ownership has not held back the companies' growth: revenues are currently around $110 billion and total employment globally approaches 800,000. The company is also extremely highly diversified, with hundreds of subsidiaries and linked firms in seventeen broad sectors including engineering, automotive, catering, hotels, airlines, financial services, information systems and communications, as well as iron and steel.

Not only do business groups comprise most of the largest firms in Asia, but this has increasingly been projected onto a global stage. As recently as 1990, among Fortune's largest 500 firms, although there were eleven from South Korea, none came from China or India, let alone the other smaller Asian economies. In contrast, by 2016, there were 125 Asian companies in the list, including 103 from China and 7 from India and those firms are almost always family, or state, owned and most are business groups. For example, India's largest single firm[6] is Reliance Industries, as we saw in Chapter 3, a business group founded in 1973 by the Ambani family. Although the company is huge by any standards – with revenues of more than $90 billion and 194,000 employees – the family retains control with Mukesh Ambani holding 47.35 per cent of the equity. Unsurprisingly, he is the richest person in India, with an estimated wealth of more than $84 billion. In China, Tencent's five founders established the company in 1998. This giant technology company, one of the top ten in the world by asset value, has an annual operating income exceeding $70 billion, and employs over 63,000 worldwide. Alibaba, the Chinese online

---

[6] The Tata group has four firms in the top twenty in India, separately listed. When combined, their turnover is larger than Reliance, at more than $100 billion. However, the Tata family is not listed in the Forbes billionaire list because much of their wealth is held in family trusts or foundations.

e-commerce and retail company founded by Jack Ma and colleagues in 1999, is of comparable size.

4.2.1 Governance of Business Groups

Asian business groups also differ from conglomerates in North America and Europe. Although the latter also have various subsidiaries, sometimes in different sectors, they follow a common strategy and typically have limited autonomy, for example with respect to investment decisions, management or strategic positioning. In Asia, business groups are usually much looser and far more flexible. The groups' affiliated companies often have their own boards of directors, management teams and even their own shareholder base. However, their controlling shareholder is either a family (or coalition of families) or the state, and the owners exercise their authority in ways and times that they see fit. For family-controlled businesses, the ties between the businesses in the group depend on high levels of personal trust that might be expected within families, as well as across families from similar ethnic or social backgrounds; quite frequently from immigrant or 'outsider' groups – such as the Chinese in Southeast Asia.[7] Where permitted, business groups often comprise a bank, suggesting that the need for capital has often been an important element in their formation.

Another defining characteristic of Asian business groups is that their corporate structures tend to be highly complex, with pyramids of stocks and numerous complicated and opaque cross-holdings.[8] This can allow them to disguise their true ownership when it suits their purposes. For example, according to a Mongabay investigation, Indonesia's Asia Pulp & Paper (APP), one of the world's biggest paper producers, has been disguising its ownership of a controversial company engaged in deforestation. APP was accused of using the names of ex-employees as proxies to hide its beneficial ownership of Muara

---

[7] Leff (1978).   [8] Claessens et al. (2006); Khanna and Yafeh (2007).

Sungai Landak, a company that profited substantially from the destruction of a massive rainforest in western Borneo. APP is also under scrutiny over its relationship with twenty-four other plantation companies linked to illegal deforestation. APP refers to these companies as 'independent', but recent NGO and media reports indicate they are owned by a handful of APP employees.

An important purpose of these pyramidal structures is to reduce the stakes required for control through judicious, indirect majority ownership of other companies. For instance, a publicly-listed firm can be controlled by a family indirectly through another publicly-listed company. To give an example; in Malaysia, the family of the entrepreneur Tan Chin Nan controls a group of companies left to them after his death in 2018. They have a 50.5 per cent stake in an entity called Goldis Berhad. The family also owns 10 per cent of the main (listed) property development company IGB in which Goldis owns a further 26 per cent. This results in the family owning 36 per cent of the voting rights in IGB Property along with 23.1 per cent of the rights to dividends. It is therefore able to exert effective control of the company while owning a lower proportion of the actual shares and, hence, at a lower cost to them. In this way, families can retain control of businesses even as they grow extremely large by slotting external shareholders into constrained holdings within pyramidal structures which they control. There is huge scope for yet more complex vehicles, for example, group affiliates owned by multiple other group affiliates, each owned by other affiliates. But whatever the complexity, the underlying purpose is to ensure that while family direct shareholdings remain quite low, the family can still exert effective control over the constituent parts of the group.

4.2.2 Pyramids and Firm Performance

Pyramidal structures and opaque governance also provide endless opportunities for self-dealing among related parties, normally members of the closely-knit family networks at the heart of the business group or their political allies and friends.[9] An important example is the transfers of resources from affiliates with high cash flows to those with low cash flows, as has been recorded in Indian and South Korean business groups. At first glance, injections of cash to struggling firms runs counter to the logic of maximising profits, under which it could be expected that resources would flow in the opposite direction. One explanation for this pattern of redistribution of profits is that the controlling family wishes to protect the integrity of the group as a whole. That is because size and diversity confer major benefits, including in the reciprocity game with politicians that we have described in Chapter 3. It may also be a part of a portfolio risk diversification strategy on the part of the owners, who therefore tend to treat affiliated companies as financial assets rather than as business organisations.

Whatever the reason, although the family can benefit from maintaining the group intact, these transfers can also be highly inefficient, as resources get transferred from higher to lower value-added uses, reducing the productivity of the group on average. This can mean that internal subsidies to inefficient from efficient parts of the group may tie up resources in poorly performing companies that may not have survived in the absence of the business group structure. Certainly, this sort of redistribution within business groups has been a widespread feature. In India, for example, the Adani Group – one with combined revenues of $13 billion and a market capitalisation that jumped from around $22 billion to over $40 billion between 2020–21 – which comprises six publicly-traded companies, has functioned in this way.[10] Adani runs its own internal capital market with a proliferation of related party transactions. Subsidiaries borrow by offering the shares they hold in the listed group companies as security. This is also used to purchase equity in sister companies, even those in unrelated businesses. For example, in 2015 Adani Properties, a subsidiary of Adani Enterprises, bought a 9.05 per cent stake in Adani

---

[9] See, for example, Morck et al. (2005)   [10] Bhaumik et al. (2017).

Transmission. Another group entity – Adani Infra – has raised a high level of loans – mostly from state-owned banks – for the group by again offering shares in group companies as security. The result of all this is that at any point of time there are multiple – and not very transparent – flows of resources between parts of the group. Some – difficult to quantify – part of this comprises resources raised from state-owned banks, allegedly secured as a result of connections.[11] As we noted in the previous chapter, the Adani Group has been singled out as one of the most politically connected to the BJP government in India and its leader, Narendra Modi. A highly visible measure of proximity was Modi's use of the Adani Group's private aircraft to fly to Delhi for swearing in as prime minister. That proximity to power was reflected in the group's market value jumping 250 per cent in the year following Modi's election.

Cross-holdings and pyramidal structures can also be used as vehicles to facilitate the expropriation of minority shareholders. Controlling families can tunnel out assets from one firm, where they formally have a minority, to another which is entirely under their control. For example, assets can be acquired from another affiliate at a lower than market price, or vice versa; or loans can be made between affiliates at above or below market rates.[12] The ability to transfer money around also helps with tax avoidance. The International Consortium of Investigative Journalists (ICIJ) found that nine of the eleven richest Indonesian families worth an estimated $36 billion – most if not all of whom own business groups – also own more than 190 offshore trusts in tropical tax havens. In 2012 the Washington-based research organisation Global Financial Integrity (GFI) estimated that Indonesia had lost over $10 billion in 'illicit financial outflows', including tax evasion, each year between 2001 and 2010. Networks play a big role here: six of the families were closely tied to the late

---

[11] Rajshekhar (2019a).
[12] Examples across Asia are well documented, for example in Cheung et al. (2006).

dictator Suharto, who helped a special circle of Indonesians grow enormously rich during his thirty-one-year rule.

To summarise the argument so far, the core of the matter is that pyramidal structures allow the dominant shareholders to exercise a level of control over their organisations which does not correspond to the level of risk that they are assuming. Concentrated cross-holdings deliver control without having to invest the appropriate money to secure it. In terms of company efficiency, the effects of such governance arrangements on managerial incentives are adverse. Rather than giving priority to profitability and efficiency, managers have to focus on asset and resource transfers across affiliates. Such practices can also stand in the way of 'creative destruction' whereby inefficient firms or units fall by the wayside and are replaced by more efficient competitors.[13] Many family firms, including business groups, particularly in subsequent generations to the founder, also tend to run into problems of nepotism, conflicts among heirs, as well as the temptations of political engagements on the back of business interests and resources.[14]

One additional consequence of the complexity of ownership formats is that it is very difficult, if not impossible, for outsiders to evaluate the finances and performance of these business groups. This lack of transparency is important in a variety of ways. Not only can it disguise the extent of internal financing between constituent parts of the group, but this opacity makes their businesses less susceptible to being targeted by politicians, including for tax raising or other purposes. This last motivation has proven to be a common response to the risk of predation; the inverse of the bounty bestowed on many of these groups by politicians in the first place.

4.3 WHY HAVE BUSINESS GROUPS FORMED: A RESPONSE TO MISSING MARKETS?

Although the examples of long-lived Asian business groups mentioned earlier are by no means unique, most business groups have

---

[13] Schumpeter (1942).   [14] Bertrand and Schoar (2006).

emerged in formats broadly similar to today at some point in the last sixty years. Yet, measuring their prevalence is by no means straightforward. In the mid-1990s the share of listed firms affiliated with business groups ranged between 37 per cent in Thailand and 73 per cent in Indonesia. In East Asia as a whole (including Japan) the average was 68 per cent.[15] A later estimate for 2005 found that 20 per cent of Chinese companies were affiliated and around 30 per cent in India. There are few, if any, reasons for thinking that these shares have declined significantly in recent years.

Many arguments have been floated – including of a governance nature that we have just discussed – as to how Asian business groups have managed to survive and prosper. One, simply put, is that it is all about trust. Low trust in external institutions leads companies to rely on family or other groups that can fulfil this criterion. The World Values Survey, which collects data across countries about self-reported measures of trust and other attitudes, shows that almost all Asian countries have relatively low to very low levels of trust, especially relative to Europe and North America. These low levels of trust in external institutions and the market will lead people to rely disproportionately on people they know personally, for example family or community members. Unsurprisingly, therefore, in Asian contexts, personal networks, such as Chinese personal networks (or *guanxi*), tend to be the foundation for business activities.[16] Others have pointed to sociological factors, such as deference to age, family and hierarchy in Asian cultures,[17] as well as the possible effect of inheritance rules. But probably the most general explanation has been the argument that business groups are, in effect, replacements for missing markets. The starting point of this argument[18] is that emerging

---

[15] Claessens et al. (2006).

[16] As an example, in Taiwan, personal connections were applied in almost every step of the founding of the Tainan Spinning group from raising capital to cultivating political connections and recruiting personnel (Numazaki, 1993).

[17] For example, Hofstede (1984) who believed that Asian societies were more collectivistic in nature than Western ones.

[18] Khanna and Yafeh (2007).

economies, such as those in Asia, have weak underlying institutions and hence higher transactions costs that in turn limit the ability to achieve efficiency through markets.

4.3.1 The 'Missing Institutions' View

Asian economies may lack many of the institutions that facilitate market transactions. Probably the most salient example concerns capital markets, which rely on complex legal contracts and hence depend on a fair, relatively cheap and effective legal system. Yet, the World Bank's Ease of Doing Business index ranks India as 163rd in the world in terms of contract enforcement. Legal claims in Delhi on average require nearly four years to be resolved at an average cost of 31 per cent of the value of the claim. Things may get resolved somewhat faster in Jakarta, Indonesia, with cases heard in just under thirteen months but average costs still amount to 74 per cent of the value of the claim. In these contexts, it is hardly surprising that business groups may be attractive because capital transactions can be kept 'within the family'. Sensitive market transactions may also be better limited to small and closely-knit groups, such as family or community.

Further, these are not the only missing markets. Managerial and highly skilled labour are also often scarce while accreditation of skills and experience is not systematic and reliable. Once again, owners may prefer to rely on family members, friends and people from their own communities. In short, transactions through imperfect or missing markets tend to be replaced by a complex skein of connections or interests in which there is trust. Additionally, it may be more attractive not to try and acquire inputs and resources, including all manner of business services, through the market but to provide them from within their own organisation. Business groups then become 'transactional arenas' that enable their affiliates to provide key inputs, whether capital or labour, while also allowing coordination across businesses.[19] This viewpoint suggests that business groups may not

---

[19] Khanna and Palepu (2000a); Khanna and Palepu (2000b)

always be a constraint on economic development, especially when market-supporting institutions are inadequate. The problems arise when they become entrenched into the economy, dominating markets and operating through the connections world to restrict competition

Thus, while these explanations of the rise of business groups can appear compelling, they fail to take account of the wider political and social context in which business groups get formed and then get perpetuated. Not only may there be more idiosyncratic reasons, such as opportunities opened up by personal connections and exchanges,[20] but it is hard to see missing markets as a convincing explanation for the very complex and opaque ownership structures of so many business groups across Asia. Nor does it really explain the close and long-term ties between politicians and these business families that were discussed in Chapter 3. These ties – we should again stress – tend to work around reciprocal favours in which supportive government regulations, competition rules, import regulations or subsidies are provided to particular businesses in exchange for supporting government initiatives and policies let alone providing bribes to, or post-retirement jobs for, politicians and their retinues. Examples abound, such as the recent case of Indonesian palm firm, Sinar Mas Agro Resources and Technology, accused of bribing parliamentarians to avoid an investigation into plantation permits and waste processing.

4.3.2 How Business Groups Evolve as Economies Develop

An important test of these ideas is therefore whether, as economies develop, business groups gradually disappear. The improvements in institutions that generally proceed in lockstep with income should render the organisation irrelevant; the benefits of internalising resources only hold when markets cannot perform the same function. Once they can, the benefits of specialisation and the division of labour should dominate. At the same time, the need for high levels of

[20] See Hamilton and Kao (1990) for Taiwan.

diversification as a way of generating sufficient resources to meet the investment needs of the group as a whole should tend to disappear as the firm is able to bring in external investors or borrow from banks. As North America and Europe became rich and institutions improved, conglomerates came to be supplanted by more focused and less-diversified organisations because they were more efficient.[21] In sum, a missing institutions perspective would lead us to expect business groups gradually to atrophy as the economy develops, being replaced by more conventional and focused corporations.

Yet despite the growth in income reported in Chapter 2 and some improvement in institutional structures, there is little, if any, evidence of a decline in the importance of business groups. One thorough study covering many Asian economies has explored this issue for a number of countries using a measure of business group prevalence, that is the number of firms affiliated with a business group as a proportion of publicly-listed firms in a country.[22] They found no consistent evidence of a decrease: indeed, in many countries there was an increase. Thus, from the late 1990s business group prevalence rose in China, Indonesia, Thailand and India, though it declined slightly in South Korea. In short, even as Asia has developed rapidly, business groups continue to be an important feature of the business landscape.

As we suggested in the previous chapter, this finding is likely to be a consequence of the way that Asian business groups have managed to entrench themselves in the political and business systems to such a degree that they are able to prevent more competitive firms and more transparent institutional arrangements from emerging. Their ability to do so depends critically on their – often long-lasting – connections to the political world. The most striking instance is undoubtedly South Korea, which is no longer an emerging economy. Indeed, it is now an OECD member, a country with income per capita already higher than Spain and ranked fifth in the World Bank's global Ease of Doing Business index. However, even here, the business group format

---

[21] Rumelt (1974).   [22] Carney et al. (2017).

has not hugely changed despite the substantial rise in income and apparent improvement in institutions.[23] Indeed, the place of Samsung and other large business groups has been consolidated.

Because Samsung has such an extraordinarily position in South Korea's economy, some refer to the company as a Republic in its own right. Founded in the 1950s, it rapidly took on a diversified business group format. It was only in the 1970s that it moved into electronics and heavy industry. Initially, the government denied Samsung entry to electronics because LG had already been selected as the 'national champion'. When allowed to enter, it was on the condition that all its products should be exported; a condition that was later abandoned. At the same time, Samsung set up a raft of high-tech manufacturing companies, often in alliance with Japanese firms such as Sanyo, NEC, Mitsui and IHI[24]. Samsung also initially benefitted from protection for its – now massive – mobile phone business.

Samsung, along with other *chaebols* such as SK, Daewoo and LG, is a reflection of South Korea's partnership with a small number of selected companies. Over long periods of time, the state has provided targeted loans and subsidies, as well as trade protection, access to foreign exchange at fixed exchange rates and designated investment projects.[25] To get a sense of the scale of support that the company has received, in 2012 it has been estimated that the company received over $150 million in subsidies.[26] These sorts of privileges in turn have boosted Samsung's growth; its sales now account for over 20 per cent of South Korea's GDP. In return for such benefits, Samsung has had a duty to maintain the state as a major stakeholder.[27] Support from, and close connections to, government have also brought entanglements of a less desirable nature. A measure of the seamier side is that more than half of the country's ten largest corporate groups are led by criminals convicted on corruption charges, including Samsung's

---

[23] Carney et al. (2017).   [24] Choi (2016).   [25] Lee and Lee (2015).
[26] One Road Research (2018).   [27] Jin (2017).

current and late chairmen; the latter – Lee Kun-hee – was also twice pardoned![28]

4.4 STATE AND BUSINESS IN ASIA

Although the scale and scope of government intervention in the economy varies widely across the rich world, most aim to transact with business at an arm's length with most business decisions influenced through non-discretionary mechanisms, such as taxation. Even when governments are more activist, the scope of their policy interventions tends to be limited to particular projects or sectors which, for one reason or another, are deemed worthy of specific attention or support. This can run as far as state ownership, although the taste for public ownership of productive assets has seriously waned since the 1980s. At the same time, while many large firms try to influence the legal and business environment in which they operate, for example through lobbying, this usually involves industry-wide standard setting, regulations or trade policies.[29]

In Asia, since 1945 and currently, the state has played a more interventionist role in the economy, including, as we have seen, through helping to form and protect the rise of business groups. Indeed, much of the success of East Asia has been attributed to the state playing a leading role in the development process, often through ownership of key elements of the capital market and natural resources, as well as the direct ownership of firms or the promotion of privately-owned national champions that benefit from massive state support. Versions of this model have been widely used throughout Asia.[30]

4.4.1 State-Owned Enterprises and Government–Business Relationships

A key pillar of this earlier drive for industrialisation through the explicit use of targeted industrial policies was SOEs. While their

---

[28] *Wall Street Journal* (2020).   [29] Keim and Hillman (2008).   [30] Kuznets (1988).

number has subsequently declined, sometimes through explicit acts of privatisation,[31] they still remain in large numbers. This, of course, is especially true in China, where currently there are still nearly 160,000 SOEs employing almost 70 million people.[32] In 2018, eleven state-owned Chinese companies had turnover in excess of $100 billion in sectors as diverse as petrochemicals, construction, automobiles, banking and insurance. In India, three of the largest five companies are state-owned or -controlled, as are four of the top five in Thailand. Vietnam's list of largest corporations is also dominated by SOEs, albeit ones with the ownership participation of Samsung.

Even when not numerically important, state-owned companies can dominate strategic areas of the economy, notably the banking sector, energy and much of infrastructure. This provides huge opportunities for governments to funnel resources to favoured enterprises and business groups, for example via soft loans from state-owned banks or pricing below cost for key inputs like electricity, construction and transport. The periodic and massive write-offs by Indian and Chinese state-owned banks testify to the highly political nature of lending and the associated weak diligence in execution.

The predilection for state ownership has, in recent decades, come to be modified. The model of large state-owned firms spearheading national efforts toward innovation, foreign investment and trade has shifted more towards supporting – through finance, subsidies and market protection – entities, often business groups, that are privately owned but agreeable to supporting key parts of the government's agenda. This echoes the national champions approach that was very successfully adopted in South Korea and, earlier still, in Japan. And, as in those cases, the lines between the state and private business groups and companies remain very blurred.

Some of the most interesting – and successful – recent cases of the state actively promoting – and in many cases financing – private companies come from China. Alibaba and Tencent are probably the

---

[31] Megginson (2017); Estrin and Pelletier (2018).   [32] Zhang (2019); Lin et al. (2020).

most visible – and successful – recent instances. Just how successful this has been is reflected in the fact that China now accounts for 55 per cent of the global e-commerce market. Let us briefly consider the mechanics of their relationship with the Chinese government.

Confident in their ability to shape the direction and – in critical areas – the independence of these companies, the Chinese government has also been relatively lax in allowing them to attract foreign capital through portfolio investment. Variable interest entity (VIE) structures have been used whereby foreign companies can obtain a degree of control over, and substantial economic interest in, certain companies without owning equity.[33] It was this model that both Alibaba and Tencent adopted, along with other major Chinese internet-based companies, such as Baidu, JD.com and Tudou.[34] However, the relationship between these companies and the government has become more tortuous over time.

Alibaba's founder – Jack Ma – started his business career in the mid-1990s but his first venture foundered as it was forced to merge with a state-owned company. The lesson was clear: fail to build close links to the Party and politicians and you won't survive, let alone flourish. With this in mind, Ma set out to build connections with many levels of government, beginning with the municipality in his hometown in Zhejiang province. The support of several so-called princelings – children or relations of Communist Party grandees, such as Alvin Jiang, grandson of former president Jiang Zemin, and Jeffrey Zeng, the son of former vice premier Zeng Pei Yan[35] – was enormously helpful, not least in sorting out how to enter the financial sector. At the same time, Alibaba has worked assiduously with some SOEs and the central government on projects explicitly favoured by government. Further, to avoid the appearance of challenging retail chains owned by branches of government, it offers a platform primarily for small brands, while the state-owned retail channels continue to host big-name products. Similarly, in the parcel service industry,

---

[33] Guo (2014, p. 569).   [34] Tang (2020).   [35] Yu (2018).

Alibaba is an integrator of local service networks without intruding into the vested interests of local operators, many of whom are tied to local government interests.[36]

Although Jack Ma is a confirmed Communist Party member, this was only revealed in 2018 and he has never played a visible political role. However, despite a long career of caution, Ma fell out spectacularly with the Chinese authorities in 2020 around the initial public offering (IPO) of his digital payment platform, the Ant Group, which was initially valued at $150 billion. Alibaba announced the planned dual listing in Hong Kong and Shanghai in July 2020, but the Chinese authorities suspended the listing two days before trading was due to commence. The full story remains unclear, but it seems that Ma had over-reached himself politically having publicly criticised both Chinese regulators and state-owned banks shortly before the aborted IPO. Ant Group itself was perhaps also seen as a threat by the authorities as it was positioned to provide finance in competition to the state banks. Whatever the reason, Ma can have been left in no doubt that he had severely overstepped the mark. New anti-trust policies targeting internet platforms were subsequently introduced in 2021 and Alibaba was fined $2.8 billion for anti-competitive practices. At the same time, the threat of competition to state banks was limited with Ant being restructured into a financial holding company. Jack Ma himself has virtually disappeared; he was last seen in public in October 2020. As he himself said, 'among the richest men in China, few have good endings'.

Tencent has also successfully used its political connections. Perhaps best known for its WeChat integrated platform, which combines access to an ever-widening range of apps and associated services, the company is actively involved in a huge range of activities across the economy.[37] Like Alibaba, Tencent has greatly benefitted from Chinese industrial policy, including the so-called Great Firewall which has kept out foreign competitors, such as Google. Even without

---

[36] Chen and Ku (2016).   [37] Jia and Winseck (2018).

explicit protection, these local e-commerce companies benefit from connections that foreign competitors, such as eBay, lack. It has also benefitted from supportive policies. For example, in 2015, the Chinese government adopted the 'Internet Plus' strategy with the aim of bringing next-generation network technologies to almost all of the sectors of the Chinese economy, including sectors that were hitherto closed or highly regulated.[38] WeChat Pay was among the first non-banking payment providers that received a license from the central bank. In 2018, WeChat Pay was allowed to launch an electronic identity card in three Chinese cities using facial recognition technology.[39] A number of local governments have also reached agreements with Tencent and Alibaba to develop smart city initiatives, linking public services such as hospital appointments or payment of utility bills with Alipay and WeChat Wallet, the two companies' payment platforms. Both companies, along with Baidu, have cultivated collaboration with some of China's state-owned behemoths.

Tencent has taken a rather different approach to the recent tightening of the leash on Chinese digital companies. Its CEO – Ma Huateng or Pony Ma – has for many years been explicitly involved in Communist Party Congresses and activities. He, along with Robin Li of Baidu, is a member of the Communist Party and the National Peoples' Congress (NPC). Even so, Tencent has also been under political pressure in the past few years, though its response has been rather different to that taken by Alibaba's founder. A voluntary meeting with anti-trust officials in March 2021 was called and at China's annual parliamentary meeting, Tencent's CEO called for stricter regulation of his own company. Despite this, Tencent was also fined in Spring 2021.

To summarise, while these companies display a rather different model of relationship between government and the private sector, aside from establishing national champions at home that are susceptible to political and strategic influence, they are also being used as

---

[38] Shen (2019).   [39] Ruangkanjanases et al. (2015).

vehicles for rapid expansion both at home and abroad. By 2016 Alibaba, Tencent and the other tech giants provided 42 per cent of Chinese venture capital (VC) investment (compared to only 5 per cent by Facebook, Amazon, Netflix and Google in the United States) and acquired 75 per cent of all successful start-up companies in the country.[40] About 20 per cent of top Chinese start-ups have been founded by the main Chinese internet companies or former employees and 30 per cent receive funding from these firms.[41] Tencent alone has also invested in forty-six global start-ups, a number comparable to Japan's SoftBank Group.[42]

What is clear is that the strategy of supporting the foundation and growth of national champions, whilst controlling some sensitive areas such as content, data privacy and parts of business strategy, has been very effective.[43] The model has also allowed scope for competition among these companies, mainly at home but also abroad. Yet, this competition opens little space for new entrants and narrows the competitive gateway. In these respects, the strategy displays a taste for market dominance and concentration that mirrors the wider appetites of business groups throughout Asia. Finally, the emergence of tensions between state and business groups in the recent past signals a broader tension between control and innovation. We will return to these critical issues very shortly.

4.5 HOW DO BUSINESS GROUPS AND SOES PERFORM?

Historically, 'state capitalism'[44] involving the funnelling of public resources to national champions, sometimes state owned, sometimes private business groups, could have been expected to help favoured companies perform better than firms that are not. Yet, the picture on this count is actually very mixed. It is also muddied by the continual problem of poor data and transparency about such companies' financials.

---

[40] Jia and Winseck (2018).   [41] McKinsey Global Institute (2017).
[42] Quartz (2019).   [43] Harwit (2017).   [44] Musacchio and Lazzarini (2014).

4.5.1 State-Owned Enterprises

With respect to SOEs, although there are undoubtedly some that have performed well in terms of profitability and productivity, the more general story is far less positive and, given widespread experience in other countries with this ownership form, hardly surprising. SOEs have almost always been found to be less efficient than private ones. Indeed, all the studies since 2005 directly comparing state-owned and private firms in emerging economies, mainly Asia, have found that performance, measured by a variety of measures including operational efficiency, productivity and profitability, is lower in state-owned firms.[45] This is usually argued to be a result of the way that the state imposes political and social objectives, as well as commercial ones, on its firms, diverting managerial attention from company performance and giving leeway for errant managers to exploit conflicting targets for their own benefit.[46] In China, where state ownership remains a significant element of the national development policy, SOEs on average remain chronically inefficient.[47] Since 2012, Chinese industrial SOEs have seen profits decline in most years, although this has not been the case for companies in private or mixed ownership. Experimentation with modified ownership and governance formats, such as state-owned holding companies, does not appear to have had any significant impact on performance.[48]

4.5.2 Business Groups

Turning to private business groups, unsurprisingly, given their pervasiveness and opacity, their impact on business and national economic performance is diverse and complex. There is a lot of material about this and Figure 4.1 summarises the argument by bringing together the possible effects over both short and long term. As we have seen, among the possible positive effects of business groups can be superior access to capital along with the creation of internal labour markets. In

---

[45] See Megginson (2016, Table 4).   [46] See Estrin et al. (2009).   [47] Lin et al. (2020).
[48] Kim and Chung (2018).

		Goods and services	Labour	Capital	Overall economy
Positive Impact	Short term	• Providing choice • Access to goods/services	• Creation of internal labour market, e.g. or managerial workers	• Creation of internal capital market	• Enhances growth when institutions weak
Positive Impact	Long term	• Development of markets	• Providing workers with insurance against downturns	• Access to funds for investment	• Becomes less diversified with economic development
Negative Impact	Short term	• Barriers to entry • Restriction of competition, supported by political connections	• Inefficient allocation of labour across uses	• Expropriation of minority shareholders • Opaque governance	• High levels of overall market concentration
Negative Impact	Long term	• Persistent rents & monopoly profits • Entrenchment of market power • Symbiotic political-business relationships	• Providing jobs in return for political favours	• Misallocation of capital, in support of political ties and networks	• Slows competition, innovation and growth • Creates great wealth inequality

FIGURE 4.1 The impact of business groups on economic performance

the longer term, their deep pockets may also ensure that workers benefit from relative job security over the business cycle. But we have also stressed clearly negative effects. Prime amongst these are lack of competition and barriers to entry, as well as inefficiencies in resource allocation and serious deficiencies in governance. Additionally, the connections web that helps entrench business groups along with a persisting lack of competition and rivalry breeds high inequality. Jobs may be created as favours to politicians and, of course, those favours can extend to outright corruption in the harvesting of public contracts and other privileges.

Commencing with firm-level performance, the obvious point of comparison is between the performance of business group affiliates relative to non-affiliates. If it is true that business groups create relatively efficient internal markets – for capital,[49] labour,[50] management and key inputs – in contexts where the market systems as a whole are unreliable, expensive or inefficient, then affiliates should be less resource constrained and as a result should perform better in

[49] For example, Almeida et al. (2015).   [50] Cestone et al. (2016).

terms, for example, of firm value and profitability. Further, it could be expected that any advantage that business groups derive for performance should decline as the economy develops.

In fact, there appear to be no substantial effects from business group affiliation on financial performance in Asia in either direction. This is not because of a lack of evidence: in fact, there are a large number of academic studies on this topic.[51] But in some countries, or in some periods, affiliates are found to perform better than non-affiliates. Other studies find exactly the opposite. This is true whether one looks at profitability or return on assets and holds for most Asian countries. A recent summary concluded that if there were any significant effects at all, they were almost always very small. For China, India and Malaysia, no differences in performance were found. However, a small positive effect was found in Hong Kong and Indonesia, and a negative one in Pakistan.[52]

How should we interpret these results? In the first place, there are bound to be serious measurement problems. Financial data about business groups are almost always opaque or unreliable because the organisations themselves are loose networks based on family ties with such information that is in the public domain mostly not transparent or comprehensive. Moreover, the tendency towards cross-subsidisation and shifting profits between subsidiaries makes it difficult to interpret the evidence as can be found on performance. For example, affiliates may appear to be performing badly relative to non-affiliates if profits are being siphoned into a family-owned private company. In that instance, one firm might appear to be performing well and the remainder poorly.

In fact, among the host of studies looking at the impact of business groups on performance, it is the evidence for a negative long-term impact through expropriation and profit or resource

---

[51] See Khanna and Rivkin (2001) and Carney et al. (2011).
[52] Keister (1998); Chang and Choi (1988); Khanna and Palepu (2000b); Khanna and Rivkin (2001); Carney et al. (2011).

shifting – often known as tunnelling – that is very strong. Pyramid-like ownership structures reduce the value of firms and harm minority shareholders.[53] To get a sense of the magnitude, in Pakistan about 15 per cent of the financial resources of affiliates were shifted or tunnelled to firms of the controlling families.[54] More generally, the financial practices – and malpractices – of business groups reduce the market value of companies belonging to a business group significantly.[55]

Yet, moving resources around among the parts of a business group is not without its benefits. Weaker parts of the group can be supported by stronger ones.[56] This may not be efficient, but it can be a very useful way of responding to a downturn or need for restructuring. But as suggested in Figure 4.1, the business group format is also very susceptible to being used for maintaining political ties and relationships and hence to perpetuating their webs of connections.[57] The accumulation of cash piles in the – often private – family-controlled vehicles at the centre of the business group web gives enormous discretion for the dominant families to distribute their favours around their networks as they see fit. These favours may extend to politicians and their preferred causes or goals, to the support of government initiatives in return for subsidies or to tax havens abroad. It can also lead to inefficiencies in the labour market, if political favours are traded for job creation in politically sensitive areas or sectors in support of the ruling elite.

4.5.3 An Explanation of the Findings

With all this in mind, the evidence as to how business groups perform is obviously not a simple story. What is very clear, however, is that business groups are organisations primarily devoted to the generation, maintenance and transfer of cash towards their controlling families,

---

[53] See, for example, Villalonga and Amit (2006); Colli and Colpan (2016).
[54] Hussain and Safdar (2018).   [55] Almeida et al. (2011).   [56] Bhaumik et al. (2017).
[57] Kim et al. (2019) find that business groups making greater corporate charitable contributions linked to the servicing of their political relationships.

and their endurance is a testament to their use of this wealth and their powerful political connections. This can happen without large costs in terms of efficiency. Indeed, when markets are highly imperfect and institutions are weak, the internalisation of resources by business groups may allow affiliates to perform markedly better than non-affiliated companies. But business groups are also very suitable vehicles for gaining and leveraging political and business connections and hence to the pursuit of rents sustained by those connections via granting licenses, giving subsidies and restricting competition, whether at home or abroad. Indeed, from this perspective, business groups are especially well-suited to the connections world of Asia. The businesses gain market power and privileges from their access to politicians and regulators, while politicians in turn can obtain leverage over large swathes of the economy merely through close association with just a few families. Business groups also offer politicians an effective route to pursue broader objectives. This tends to involve not just the provision of preferential finance and contracts but also inhibiting competition from international rivals, an approach followed with some success in China by Alibaba and Tencent among others.

Although most Asian business groups are focused on domestic markets where their connections are particularly potent and helpful to them, the international arena can be a far tougher place, not least due to competition from multinationals. Yet, some Asian business groups have expanded overseas, sometimes, but by no means always, with active support from their governments. Such support has included export subsidies along with facilitating technology transfer from advanced economies, as through special economic zones. Entry into foreign markets has also led business groups to invest in improving their own performance in order to compete effectively. Examples include Lenovo or Haier in China, Acer in Taiwan and Samsung in South Korea.[58] India's established business groups, such as Tata, have

[58] Lamin (2013).

deliberately developed marketing and technological competences that have fuelled their investment abroad and growing role in markets outside their home countries.[59]

Thus, concentrating on business groups' performance whilst important, if often elusive, risks camouflaging the main purpose of these entities, namely, their perpetuation and that of the family interests that lie behind them. This often involves behaving in ways that are, among other things, prejudicial to minority shareholders and rely on a pervasive lack of financial and other transparency. As a consequence, it is hardly surprising that performance varies significantly. Business group affiliates can have stronger performance because they have access to more resources along with market power, or they may perform worse because their cash is siphoned out to serve the purposes of their owners. In a sense, performance is not really an appropriate measure. That is because the survival and growth of business groups is primarily about wealthy and well-connected families using and shifting around resources according to their own purposes. This may be inefficient, as when they prop up ailing businesses, or even corrupt. But it may also improve efficiency, as when they lead the advance onto global markets. What is absolutely clear is that it always involves the accumulation of power, cash and connections in relatively few hands. And, perhaps most saliently, it involves the accumulation of market power.

4.6 BUSINESS GROUPS AND MARKET POWER

A characteristic of business groups – particularly in the longer term – is that they almost automatically erect barriers to entry for potential competitors, not least because of the size and scope of their activity, a feature already signalled in Figure 4.1.[60] Preferential access to resources or favourably-priced inputs, such as for electricity, along with internal capital markets that give them access to financing typically not available to non-affiliated competitors, are important

---

[59] Siegel and Choudhury (2012); Chari (2013). [60] Pattnaik et al. (2018).

sources of advantage. These barriers can be raised further in sectors where regulatory actions really matter – such as telecoms or infrastructure – by business groups' proximity to power, especially when the former are framed as national champions spearheading economic development. Further, trade impediments have commonly been placed in the way of foreign competitors, to the benefit of Alibaba and Tencent, among others, in China, but also for Samsung in South Korea, and major local business groups in the Philippines and Indonesia as well as in other countries.

There is already much evidence that markets are more concentrated in emerging compared to advanced economies.[61] To get a good vantage point on the extent of concentration and the place of business groups, we need now to look at how markets are currently arranged. In an influential book, Thomas Philippon (2019) has argued that since 2000, concentration has been rising across US industries and increased barriers have limited the entry of new competitors. An important consequence has been an increase in incumbent companies' profit margins and a reduction in the share of labour in national income. Asian structures of business and the wider institutional arrangements that this chapter has described make for an even more potent brew from the same ingredients. Indeed, what has happened can best be summarised in terms of economic prevalence. To see how this is the case can best be captured by focusing on the scale of business group activities relative to those of the economy as a whole. This is measured through a five or ten firm concentration ratio (CR). These ratios (CR5 and CR10) represent the total revenue of the largest five (ten) firms in a country as a percentage of that country's GDP.[62]

---

[61] Tybout (2000).

[62] The measure is clearly imperfect because revenue is a sales concept and GDP a value added one. Thus, the measure can vary across countries with the share of revenue in value added. Moreover, sales include domestic and foreign sales, and highly international firms may have a smaller domestic impact. Even so, the measure gives an interesting starting point for measuring economic power in a

4.6.1 Measures of Overall Concentration

To make this calculation, we have collected information on the revenues of the largest firms in Asia. Since some of these are not listed, either because they are state owned or because they are private, we calculate separately the economic prevalence levels of both the largest listed companies, as well as for all companies including unlisted firms and SOEs. Including unlisted firms increases concentration in a few countries where state or private ownership of very large firms is common, notably Vietnam and, to a lesser extent, China.

The results are reported in Table 4.1 and are for 2018. Economic prevalence is indeed very striking. In four economies, the share of the top ten firms (CR10) – listed and unlisted – exceeds 25 per cent of GDP. These are extraordinarily high levels of concentration. Put starkly, in many Asian economies, half a dozen or fewer organisations generate revenues that constitute much of the country's economic activity. Unsurprisingly, economic prevalence rates are found to be somewhat lower in countries such as China, India and Indonesia because their economies are so much larger and more geographically and politically diversified. This means that fewer firms are able to cover all provinces or states and there are geographic barriers to unifying markets, allowing a greater localisation of business. Even so, the largest ten listed, as well as unlisted and state-owned, behemoths account for more than 15 per cent of GDP in China and India, with the top five generating more than 10 per cent. In both Pakistan and Bangladesh, countries with relatively low-income levels, significant agricultural sectors and historical factors that have limited the ability of firms to achieve scale,[63] CRs are significantly lower at below 10 per cent.

country wielded by the largest firms. The table is constructed using Orbis and cross-checked using Fortune 500 and country sources.

[63] Nahid et al. (2019). In Bangladesh, much employment, even in large companies, is indirect or outsourced. We return to this issue in Chapter 6.

Table 4.1 *Economic prevalence in Asia (five and ten firm revenue–GDP concentration ratios)*

Country	Unlisted & Listed CR5 (%)	Listed CR5 (%)	Unlisted & Listed CR10 (%)	Listed CR10 (%)
Bangladesh	3	3	4	4
China	11	9	16	13
India	11	11	17	16
Indonesia	4	4	7	7
Malaysia	11	10	18	16
Pakistan	6	5	8	7
Philippines	19	19	27	27
South Korea	30	30	43	43
Thailand	27	27	40	36
Vietnam	36	10	46	15

Source: Authors' calculations

Our measures probably understate the extent of economic prevalence. This is because the data for the largest business groups cannot always be consolidated. For example, in India, four of the ten largest companies are members of the Tata group, which does not report consolidated accounts. Unfortunately, in such instances there is no simple method that allows us to compute the actual total business group revenue in a consistent and reliable way. But we can get an idea of the extent of underestimation. For Indonesia, we calculated the CR when every firm in the country's top fifty companies was consolidated into business groups using information about the firm's ultimate owner. When doing this, the CR5 ratio for Indonesia increased by more than 25 per cent. Adding more, and smaller, firms outside the top fifty, the share would doubtless have grown further.

To put these numbers in wider context, the CRs are about 50 per cent higher than comparable measures for the United States. Put bluntly, Asia is marked by an extraordinary concentration of economic power in rather few hands. This process of economic prevalence has been achieved on the basis of family-owned business groups, along with state-owned firms. Both organisational formats have maintained close formal and informal relationships to power and the political class. The common result is that the extent of competition and rivalry has been materially reduced. As we will argue in the next chapter, some of the longer-run consequences of this concentration are likely to be lower innovation and productivity growth for the economies as a whole even as the profitability of incumbents rises.

4.7 STAGGERING ACCUMULATIONS OF WEALTH

Another way of capturing economic prevalence is to look at what has happened to private wealth in Asia. As was indicated in Chapter 3, high levels of market concentration driven by, or supported through, the connections world are highly supportive of large-scale accumulation of private wealth. Table 4.2 shows the number of Asian billionaires reported in the Forbes lists for 2000, 2008 and 2020, as well as their average and median wealth. The growth in the number of billionaires in Asia has been quite staggering – from only 47 in 2000 to 719 twenty years later. This is mainly because of China. In 2000 there were probably no, or very few, billionaires in China (data are not available) but by 2008 there were forty-two, second only to India's fifty-three billionaires. This number then jumped more than nine-fold in the next twelve years. By 2020, more than half of all billionaires in Asia were located in China and most are the owners of businesses, one consequence of which has been that the share of inherited wealth relative to self-made wealth has fallen dramatically (a feature also of South Asia).[64] It seems that the current leadership of China has taken to heart, if not to excess, Deng Xioping's alleged call to arms: 'to get rich is glorious'. Much has been written of the way that China has been able to pull hundreds of millions of people out of poverty.

[64] Freund (2016).

Table 4.2 Billionaires in Asia and their wealth (US$)

Country	2000 Number of billionaires	2000 Average wealth (US$bn)	2000 Median wealth (US$bn)	2008 Number of billionaires	2008 Average wealth (US$bn)	2008 Median wealth (US$bn)	2020 Number of billionaires	2020 Average wealth (US$bn)	2020 Median wealth (US$bn)
China	n/a	n/a	n/a	42	2.0	1.5	389	3.0	2.1
India	9	4.3	2.9	53	6.3	2.5	102	3.1	1.8
Indonesia	2	2.1	2.1	5	2.2	2.0	15	3.3	2.2
Malaysia	5	2.5	1.8	8	4.1	2.8	13	3.6	2.6
Philippines	5	1.5	1.5	2	1.5	1.5	18	2.5	1.9
South Korea	1	2.8	2.8	11	1.8	1.8	28	2.5	1.5
Thailand	1	1.2	1.2	3	2.9	3.6	20	3.3	2.1
Vietnam	n/a	n/a	n/a	n/a	n/a	n/a	4	2.6	1.8

Source: Forbes and authors' calculations. Empty cells imply missing information.

The other side of the coin – the creation of extreme wealth – is less well appreciated or even discussed.[65]

The growth of extreme wealth is not just about China. The rise in the number of billionaires in the past twenty years was more than ten-fold in India, South Korea and Thailand, and more than five-fold in Indonesia. Further, the average net worth of billionaires increased substantially, for example, by 175 per cent in Thailand, 71 per cent in Indonesia and 67 per cent in Philippines in the same period. In China, average wealth also rose by 50 per cent between 2008 and 2020. In short, the growth of private wealth in Asia has occurred through the massive creation of new billionaires. Moreover, the average and median wealth of billionaires has actually risen in almost all countries, despite the rapid increase in their numbers. The same story of business groups and government connections lies behind many of these ascents.

It is not only about the creation of relatively new wealth. At the same time, existing billionaires have mostly been getting richer. Although the massive surge of new billionaires in India forced down average wealth between 2000 and 2020, for the seven billionaire families that were in these lists in both years, their wealth on average increased by around 380 per cent. This includes several of the dynasties we introduced in the previous chapter, such as the Ambani family, as well as the Birla's, whose business group with $50 billion in revenues remains one of the largest in India. The Hindujas – also owners of a top twenty business group – experienced an eleven-fold increase in wealth. The story is repeated across much of Asia, even if at a lower intensity. South Korea's only billionaire in 2000, Lee Kun-hee, the ex-chairman of Samsung, saw his family wealth jump from $2.1 to $28 billion by 2020, while the Zobel family in the Philippines, owners of several major business groups, saw their wealth climb from $1.3 to $6.3 billion. As we would expect, the families

---

[65] Although, to be fair, rising inequality of income has been quite widely noted, as in Milanovic (2016) and Piketty et al. (2019).

controlling Asian business groups dominate the list of billionaires in every country.

4.8 CONCLUSION: THE CONSEQUENCES OF CONCENTRATION

Pulling these many strands together, it is evident that modern-day Asian capitalism is characterised by well-entrenched, large, family-dominated business groups. Most have very non-transparent forms of governance and have been able to carve out significant market power at the expense of a more competitive economy. In some countries, these are complemented by large state-owned companies, although increasingly governments have shifted in favour of privately-held but state-supported champions. Whatever the country level mix, the consequences have been rather similar. Our measures of economic prevalence are really very high even when compared to other countries that have experienced reductions in competition. In Asia, the consequence has been economic dominance of the economy and the policy environment by a relatively small group of firms, interconnected across product and factor markets through formal and informal structures of ownership. The wealth of the owners of these business groups has continued to spiral. Much of this has required the consent of power and has involved complex and varied forms of collusion between politicians, political parties and connected businesses.

A central conclusion that needs to be drawn is that it really makes little sense to try and separate out the existence and impact of business groups from the wider connections world that is characteristic of the Asian countries. Business groups exist for a reason. Although the format may help them address missing markets, at the same time, it also permits them to achieve dominant positions in markets by exploiting ties to the world of politics. Markets for capital and labour have undoubtedly emerged as Asia has developed, yet business groups, and the concentrations of wealth that they generate, have persisted and, often, increased their entrenchment. In part, this is because many of these business groups have had the advantage of

being first movers – itself generally a consequence of their connections – but it also reflects the long-term trend towards entrenchment
and high levels of market concentration that has been associated with
the symbiotic relationship between controlling families and the political class in most Asian countries.

As we have already noted, many of the consequences of this
symbiosis are clearly not desirable. Business groups perform in ways
that are prejudicial to minority shareholders, allow tunnelling and do
not provide financial and other transparency. In terms of their
accounting performance, business group affiliates may perform better
because they have access to more resources, as well as greater market
power, or they may perform worse because their cash is siphoned out
to serve the purposes of their owners. To emphasise, the main impact
of business groups is therefore not best indicated by their profitability.
Rather the long-term survival and growth of business groups is primarily about wealthy and well-connected families shifting very considerable financial resources around their economies according to
their own purposes. This may be inefficient, as when they prop up
ailing businesses, support politicians' agendas or merely implement
trophy projects. It may even be corrupt, powered through the incestu-
ous government–business links and networks that we have discussed
in detail. But it may also be efficiency enhancing, as when they lead
the advance onto global markets supported by their governments
using strategic firms as so-called national champions. But undoubtedly the most significant impact of business groups and one that cuts
across financial performance is the way they accrete market power
and deter entry. The disproportional dominance of a few firms in
Asian economies, and the associated concentration of family wealth,
illustrates this very starkly.

The accumulation and entrenchment of massive power, cash
and connections in a few hands will be a major barrier to broad-based
growth and development in Asia in the future. Yet the barriers to
changing this configuration – to which we will return in Chapter 7 –
are substantial. It is not just that incumbents aim to thwart attempts

to reduce or eliminate the advantages they possess but that the politicians who benefit formally and informally from their existence also have powerful vested interests in their preservation. This makes for a powerful coalition of incumbents with the power and resources to entrench their positions into the long term. And as we have seen, the wealth to be protected by billionaire families, their dependents, their retinues of politicians and their networks is staggeringly large. One wider consequence is that a process of creative destruction whereby new more efficient firms replace older, less efficient ones – a hallmark of competitive capitalism – will be far weaker. This will tend to impede the growth of productivity and, by raising entry barriers and restricting competition, limit innovation, issues that we consider further in the coming two chapters.

5 What Scope for Innovation?

5.1 INTRODUCTION

Whatever the problems Asian economies may be storing up for themselves because of the connections world, this is not preventing commentators from waxing lyrical about Asian, and especially Chinese, innovation. Consider the following:

> As an emerging tech giant, China has demonstrated it can be a leading innovator both globally and domestically. The country is making gains in ...: 1) manufacturing, 2) digital platforms and associated markets; 3) the utilization of apps and other technologies designed 'to solve societal problems' and reconfigure existing businesses in the process; and 4) basic science R&D in fields such as computing and biotechnology.
>
> *(Schoff and Ito, 2019)*

> Chinese companies have surged ahead of their U.S. counterparts on a Nikkei ranking of the top 50 patent filers for AI over the past three years.
>
> *(Nikkei Asia 2019a)*

> Today, every senior executive of a Western corporation needs to understand the tidal wave of innovation flowing from China that is about to engulf Western markets.
>
> *(Yip and McKern, 2016, p. 3)*

Indeed, newspapers, think tanks and multilateral agencies commonly highlight Asian advances – software and pharmaceuticals in India;[1]

---

[1] According to Equitymaster 22 January 2020, 'India is the leading (software) sourcing destination across the world, accounting for approximately 55% market share'.

electronics and mobility in South Korea; AI, fintech[2] and renewable energy in China. Even the more sober World Bank, argues that 'China has built a large and extensive National Innovation System to supply the innovation and technologies required for productivity growth, making it the first middle-income country to join the 20 most innovative countries'.[3]

These and other – often rather breathless – books and reports leave a strong impression that, whatever the structural deficiencies identified in this book, some, if not all, of Asia is already drawing on innovation to power itself forward. But such assertions aside, are the Asian economies really that well placed to make the transition to a growth process driven by innovation, especially considering the entrenchment of the connections world?

To come to grips with this big question, we first need to clarify what exactly is meant by innovation along with what drives it and the role of different types of organisations – stretching from SOEs and business group behemoths to small, new entrepreneurial ventures – in the innovation process. Most importantly, how effectively can the delicate task of nurturing ideas from the laboratory to the consumer be achieved in the connections world? Do the benefits of incumbency and entrenchment give Asian business groups and SOEs the security and financial muscle to dare to innovate? Do the advantages of concentrated resources offset the benefits of experimentation and spreading of risks? Or does the world of connections blunt incentives for change, while simultaneously blocking outsiders from using innovation to obtain a foothold on the ladder? These are the themes to which this chapter is devoted.

As will become clear, the evidence so far from most Asian economies is that innovation remains fairly limited and that outside a few sectors, the proliferation of small, innovative companies – a

---

[2] For example, *The Economist* (2020b), 8 October, refers to China's pre-eminence in digital money.

[3] World Bank Group and Development Research Centre of the State Council, P. R. China (2019).

hallmark of innovation elsewhere – is limited, held, as it is, in check by the hegemony of business groups. Even so, when thinking about a country's potential to innovate, the size of an economy can clearly matter. Because of that, particular attention has to be paid to China and India. In the former, there is the political imperative to become more innovative, now made keener by increasing tensions with the United States along with the government's desire to don the mantle of a global superpower. In the latter, remarkable progress in the ICT and pharmaceutical sectors, along, more recently, with strides forward in fintech, has sometimes been taken to presage innovation on a wider front.

5.2 INNOVATION'S ROLE IN DEVELOPMENT AND WHAT DRIVES IT? A HISTORICAL PERSPECTIVE

Innovation, growth and productivity are inextricably linked. Most economists view innovation as a central driver of economic growth, at least in developed economies. Robert Solow, Nobel Prize winner and the father of modern economic growth theory, argued in his 1957 paper that economic growth is driven by both increases in factor inputs – labour and capital – and what he called technical progress. The latter is measured by increases in labour productivity and is driven by the process of scientific advance, invention and innovation. Solow[4] found the accumulation of capital and increasing employment only explained around 12 per cent of the growth in labour productivity in the United States between 1909 and 1949. The remainder, sometimes called Solow's residual, was explained by technical progress, which is driven by innovation. This seemed also to be the case for Britain in the eighteenth and nineteenth centuries.[5] Science, invention and new production processes, brought to market by entrepreneurs, are therefore at the heart of economic growth in advanced economies.[6]

[4] Solow (1957).   [5] Mokyr (1990).   [6] See Ridley (2020).

But this is by no means universally the case. The so-called East Asian economic miracle[7] between 1965 and 1990 and, before that, the rapid economic development of the Soviet Union from the 1930s onwards, shows that not all, or even most, growth needs come from innovation. These various economies developed mainly through extensive growth involving the application of more and better educated labour, along with increasing quantities of capital in the industrial sector and usually in cities. This mainly took place using established, rather than new, production techniques and mostly applied to the manufacture of existing, instead of new or innovative, products. In other words, economies can grow with only limited amounts of innovation, especially if they are catching up with technologically more advanced countries.

5.2.1   Innovation and the Stages of Economic Growth

The role of innovation in driving growth will thus depend on the level of development of each economy. Acknowledging this, Michael Porter distinguished three different stages of economic growth.[8] In the first stage, when economies are *factor-driven*, there is a reliance on agriculture, natural resources and the manufacture of handicrafts and low value-added goods. Countries at this stage create little scientific knowledge as a basis for innovation.

In the second *efficiency-driven* stage, the economy grows primarily by transferring labour from low productivity uses in agriculture to higher productivity uses in manufacturing. For this to happen requires both the accumulation of capital to purchase machines that raise productivity and better education and skills so that workers are able to use the new machinery at their disposal. In the United Kingdom in the eighteenth century, the process of innovation, entrepreneurship and mechanisation were closely intertwined, but in the modern world, countries in the second stage do not necessarily need

---

[7] See Baily et al. (1998).   [8] Porter (1990).

to innovate themselves. They often rely instead on innovations from elsewhere, adapted to their own conditions.

The third stage of economic growth is *innovation-driven*. When the possibilities for greater or more intensive use of labour and capital have been largely exhausted, countries must mainly rely on innovation to generate growth, as in Solow's analysis. This implies reliance on the invention of new technologies, processes and products, which itself depends critically on the spread and quality of higher education, especially the science base in a country.

What this framework makes clear is that the form of innovation in each country, and the relationship between innovation, productivity and growth, will depend upon the stage of development in which the country finds itself. Advanced economies, such as the United States, Germany or Japan, are mainly in the innovation-stage. Growth depends on a country's capacity for, and effectiveness of, innovation. This is well understood by policymakers. For example, the European Union explains its lacklustre growth performance in recent years by deficiencies in invention resulting from a scientific brain drain; lower levels of R&D (0.8 per cent of GDP less than the United States and 1.5 per cent less than Japan); and a policy environment which is insufficiently 'innovation friendly'.

In contrast, countries in the efficiency-stage – as applies across much of Asia – depend for growth on applying increasing amounts of labour and capital and using established technologies. What role does innovation play in the efficiency-driven stage of development? At first sight, it may seem to be rather limited. Such economies do not need to invent new products or generate their own original production methods, obtaining technology instead from advanced economies. Rather, their attention is focussed on transferring labour to higher productivity jobs, improving labour skills, and accumulating capital and infrastructure. Consequently, they do not need to rely to a significant extent for their economic success on their own science-based innovation of new products or processes as the principal driver of productivity growth.[9] Even so, this still leaves them with major innovation problems. For example, they need to find ways to ensure a smooth and effective process of *diffusion* of existing technologies from the advanced economies. Further, their local specifics – such as consumer preferences and institutional arrangements – may mean global products and processes need to be *adapted* significantly for local contexts.

5.2.2 *Innovation Is Not Invention*

When we think about the relationship between an innovation and growth, there is an important distinction to be drawn between *inventions* – the development of new products through scientific research processes – and *innovations* – their successful introduction to the market place.[10] To be successful at invention, a country needs a strong scientific and engineering establishment, built around high levels of education and an open, absorptive research culture. Invention therefore tends to be located within research networks centred around leading universities, although it can often be financed, and sometimes undertaken, by private firms in their own laboratories as well.[11] In contrast, innovation is primarily a business activity. While it also benefits from a well-educated population, the critical talents relate more to business acumen, understanding the market and assembling key resources, notably in design, finance and a skilled management team. Although both invention and innovation tend to

---

[9] They may also be impeded by the migration of creative individuals to advanced economies. However, such migration is not necessarily a brain drain; see Commander et al. (2004).

[10] Crossan and Apaydin (2010) provide a widely used definition of innovation: It is 'production or adoption, assimilation, and exploitation of a value-added novelty in economic and social spheres; renewal and enlargement of products, services, and markets; development of new methods of production; and the establishment of new management systems. It is both a process and an outcome'.

[11] The two approaches are epitomised in the development of vaccines to counter COVID-19. Some, such as AstraZeneca were developed in university labs and produced by a private firm. Others, such as Pfizer, were developed by a private science company and produced by a private firm. Some, such as Johnson & Johnson, were developed and produced entirely by a private firm.

flourish more in environments with free and open supportive networks, for the former these are primarily communities of scientific scholars while for the latter, the supportive ecosystem also includes serial entrepreneurs, business angels, venture capitalists and property rights lawyers.

Take the example of the invention of graphene – a single layer of carbon arranged in a two-dimensional honeycomb lattice. It had been theorised about for many years but was only first observed in 1962 and isolated in 2004, winning Andre Geim and Konstantin Novoselov from the University of Manchester the Nobel Prize in Physics in 2010. The wide range of potential commercial applications of graphene are only now being identified and exploited, for example batteries, solar panels, lubricants, paints and 3D printers. As such, a particular innovation may represent one of the many alternative applications of scientific inventions. The application can be a new and more efficient way of making something, such as the Bessemer steel process or producing glass using the float method. This is usually referred to as process innovation. But it can also be a new product, such as an automobile, airplane or a mobile phone. This is known as product innovation.[12]

In sum, scientific inventions may provide the basis for innovations, but the relationship is not linear. This is partly because a single invention rarely leads to a single innovation. Rather, innovation proceeds in a more serendipitous way by combining and recombining a variety of new and not-so-new scientific inventions. The rise of the railways came from advances over many years in steel manufacture, compression, pistons and the science of gases, to name but a few of the underlying scientific advances. Even so, a strong scientific base and an educated labour force, including of engineers able to draw practical applications from abstract scientific theories, are usually

---

[12] Innovations can also be less path breaking than these examples, such as a better, cheaper and more attractive way of doing the same thing. For example, the gradual replacement of incandescent light bulbs, developed in the 1920s, by LEDs and halogen light bulbs to save energy.

considered to be a necessary, if not sufficient, condition for countries to innovate successfully.[13]

5.2.3 Radical and Incremental Innovation

It is also useful to apply a further categorisation that indicates how disruptive the innovation is to existing production processes or consumer behaviour. We can distinguish between innovations that are *radical*, and those which are *incremental* (or gradual). Both will be important for growth in the innovation stage of development. Radical innovations are what people usually think about when they use the term 'innovation' – they transform markets and consumer behaviour. Think of the development of the telephone, the radio, television or the Internet. For processes, think of the application of robots and AI to manufacturing and increasingly to services – the so-called Fourth Industrial Revolution.[14] Another example is the impact of the widespread use of the container for transport of goods, logistics and ports, which drastically reduced the costs of transport of traded goods.[15]

One of the most important characteristics of radical innovations is the way that they disrupt existing relationships between producers, as well as between consumers and producers, thereby opening up the possibility for new firms to enter what were previously mature markets.[16] Standard examples include how Amazon used the Internet to subvert the dominance of the high street and other retail outlets, and how the growing policy emphasis on electric vehicles created a market opening for Tesla. Radical innovation is often associated with new firm entry and industry *disruption*, whereby existing incumbent firms are threatened and often replaced by new companies.

In contrast, incremental innovations draw on a deep understanding of specific processes and products, as well as how buyers want products to evolve. Since Henry Ford introduced mass production of the internal combustion engine vehicle, the car has only

---

[13] Fagerberg et al. (2005).   [14] See Schwab (2016).   [15] See Levinson (2016).
[16] Christensen et al. (2015).

gradually advanced in every dimension: design, performance, fuel efficiency, internal comfort, application of electronics and communicability. These are incremental innovations and have rarely, if ever, been disruptive. Incremental innovations tend not to generate a huge amount of entry or threaten incumbent firms.

The distinction between radical and gradual innovations is important because, while the latter are often developed by existing firms, radical innovations often represent the fruits of individual or group innovative activity outside the domain of current organisations. Moreover, it is here that agency enters our story. Tales of innovation, especially radical innovation, usually have as their central character an individual – the entrepreneur.

5.2.4 Entrepreneurs and Innovation

The entrepreneur is the individual (or group of individuals) who brings new processes or products to market, often by commercialising scientific inventions or combining new and old inventions. The Austrian economist, Joseph Schumpeter, was the first to identify the key role of entrepreneurs in national development.[17] The entrepreneurs' essential strength is not necessarily with the science itself, though knowledge in this area can be important. After all, Bill Gates needed to understand programming and Steve Jobs needed to understand computer hardware. But for both of these individuals, and for countless other entrepreneurs, their key contribution is identifying the gap in the market and designing a product that fills it and at the right price.

This does not happen overnight. Countless examples attest to the role of trial and error as entrepreneurs gradually hone their product to meet demand that consumers did not previously know that they had. Often this experimentation costs rather more than the available financing, so the entrepreneur may well be bankrupted before any radical innovation bears fruit. In such cases, the benefits

---

[17] Schumpeter (1934).

are normally picked up subsequently by others.[18] Unsurprisingly, therefore, entrepreneurs have to be resilient and have an appetite for risk. Moreover, inherent to the idea of entrepreneurship is the creation of a new organisation: the entrepreneurial firm.[19]

5.2.5 Incumbent Firms and Innovation

Existing firms, including business groups, can, of course, also be innovative. If we think about Sony from the 1960s to the 1990s, we see the laboratories of one company producing innovation after innovation from the Walkman to the Discman. Similarly, the labs of Apple took us from personal computers to iPods to iPhones. Steve Jobs himself popularised the term 'intrapreneurship' to describe entrepreneurial activities taking place within the firm.[20]

Indeed, incumbents have many advantages in the innovation process. For existing products and processes they already understand the market, consumers and underlying technology. Business groups in particular often have strong internal funding along with the appropriate human resources to devote to innovation. All this points to existing firms as a natural location for successful innovation, especially gradual innovation.

Nonetheless, there are counterarguments. Consider the development of electric vehicles (EVs). We might have thought that existing auto manufacturers, with their expertise in managing supply chains, ensuring quality, brands and marketing, would be the natural places to develop a new type of engine for a vehicle. However, for decades, existing manufacturers did not make major advances in this field and have now only begun to do so, spurred by intense competition from a novel entrepreneurial entrant, Tesla, led by serial entrepreneur Elon Musk. Even so, now that the market prospects for EVs have been established, it could have been expected that the existing manufacturers – such as Ford, Volkswagen, Toyota and

---

[18] There are numerous examples of these processes in Ridley (2020).
[19] Davidsson (2016).   [20] In Lubenow (1985).

Daimler – would have exploited the advantages of incumbency and Tesla's inexperience in mass producing cars of high quality at low cost. But despite these apparent advantages, the incumbent's' advantages have yet to be realised.[21] Indeed, the capital markets have given their own verdict: in 2020, their valuation of Tesla was larger than the top five traditional auto producers put together!

This argument is not specific to the car industry. Existing firms often find it difficult to respond to the challenges created by innovation in their markets. You only need to think of Xerox in photocopying after its patent expired or Kodak's inability to exploit its huge advantage in photography when digital cameras appeared or Sony's difficulty in adjusting to personal computing and mobility. An important part of the reason for this is that management does not want to kill the goose that lays the golden egg! If a firm makes the bulk of its profit from a particular product or service, it will often face deep organisational and incentive problems when developing a new product, especially one that will eat into, and ultimately displace, its existing product lines.[22]

The central role of experimentation represents a further problem for entrepreneurship in existing firms (sometimes known as intrapreneurship). Endless experience shows that it is a complex matter to turn new ideas into products that consumers are willing to buy at prices at which it is profitable. In fact, many if not most new products initially fail, or take several attempts to become established. Yet, large corporations rarely have a culture tolerant of failure. Moreover, successful innovation can take a long time. Uber has grown very rapidly since its foundation in 2009 but it has not yet posted positive profits. Corporations tend to march to the tune of stock

---

[21] Estrin et al. (2020).

[22] The problem is not insoluble, of course. Firms have had success by forming 'creative' units well isolated from the main corporate structure. Skunk Works unit at Lockheed Martin was the most famous example. Others include Gmail invented by Paul Buchheit at Google and Spencer Silver and Art Fry who invented Post-It notes at 3M.

markets with quite short-term perspectives. Generally, the trial-and-error processes associated with innovation usually require more intellectual freedom and organisational flexibility than is common in existing firms.

Because of the key role of trial and error in innovation, new, small, entrepreneurial firms will necessarily be amongst its most significant drivers.[23] Highly centralised organisational models may be capable of advancing science for large and focussed projects, for example, getting a man on the moon or the Manhattan project in the United States. However, they tend to be much less effective at ensuring the transfer from scientific laboratory to profitable business model, or to adapting products until they satisfy consumers in different contexts. For these, a process of trial and error tends to be much more efficient. This also helps explain why innovation is often associated with an explosion of new firms, many of which will ultimately fail.

5.2.6   The Role of Entrepreneurial Ecosystems

Successful innovation, and perhaps especially radical innovation, therefore, has to be built upon a rich environment of scientific advances combined with mechanisms that allow these advances to spill over into commercial applications. Success is not only about a few world-class scientists and outstanding laboratories, although they are relevant. What is actually required is an *ecosystem* of universities, scientists, engineers as well as lawyers, financiers, skilled workers and entrepreneurs to turn a plethora of abstract scientific developments over a variety of fields of research into a stream of new products.[24] There are also important agglomeration effects: all of this works better when ecosystems are larger.[25] Moreover, because innovation is inherently uncertain – one cannot know in advance exactly what will be developed and what will succeed on the market – the

---

[23] This is an argument made persuasively by Audretsch (1995).
[24] Mason and Brown (2014).   [25] See Jacobs (1961); Audretsch and Feldman (1996).

process is most effective in environments in which large numbers of people are gathered and, most importantly, where they can communicate developments in their own field to others working on parallel problems. This leads to virtuous cycles of cross fertilisation.

Silicon Valley – centred around some of the greatest scientific universities in the world and filled with engineers and businesspeople combining and recombining science and ideas, watched, encouraged and financed by business angels, venture capitalists and patent lawyers – is the most striking example and model of decentralised innovation. Silicon Valley exists as an ecosystem. It provides an answer to the question – if small firms are more innovative than large ones, from where do they get their knowledge and the funding to develop their ideas? This is especially important given that large firms usually hold the cards in terms of the resources needed to finance the R&D from which much innovation will ultimately derive. Silicon Valley ended up as a cluster of skills, business services and financiers formed in an ad hoc way.[26] When scientists, engineers, MBAs and financiers are co-located in localised areas, meeting in coffee shops and cafes, this helps generate spillovers of knowledge within and across fields to stimulate innovation particularly in knowledge-intensive industries and earlier stages of an industry life cycle. Co-location has also led to the development of new forms of capital market institutions specifically catering to the high-risk, high-return experimentation cycle of high-tech industries. These are, notably, venture capital (VC) firms but also private equity and angel investors.[27]

This may have all started in California, but most advanced countries now have developed ecosystems of this sort, clustered around universities and large cities that contain supporting financial, legal and business services.[28] Examples include London, Berlin,

[26] These ideas were developed by Saxenian (1994).
[27] The rise of the venture capital industry is analysed by Gompers and Lerner (2001).
[28] See O'Connor et al. (2017).

Boston and Tel-Aviv. The crucial ingredients include a mix of complementary skills, supportive political structures and a robust property rights system, especially for intellectual property rights.

In sum, different countries have different 'national innovation systems'.[29] There are multiple models to success. Among developed economies, the so-called Anglo-Saxon countries tend to focus on radical innovation with their efforts centred around universities and rich research environments supported by privately-organised, highly competitive and well-financed ecosystems. Countries such as Germany and Japan – also major innovators – have placed more emphasis on incremental innovation with much of the effort centred on publicly funded laboratories with close links to large corporations who tend to provide the relevant ecosystems internally.

5.3   INNOVATION IN ASIA TODAY

With the scene set, how do we evaluate the performance of Asian economies today in terms of innovation? The obvious place to start is to consider the science base in terms of skills or human capital along with R&D expenditure. In most Asian countries, as would be expected in countries in the efficiency-driven stage of development, these are relatively weak albeit with considerable, recent progress in a few countries. As we shall see, much of Asia's innovative activity has taken place in three countries: China, India and South Korea.

5.3.1   R&D and Invention

Table 5.1 summarises the available evidence on the science base and innovation output of the main Asian economies. Looking at R&D – and bearing in mind that the share of R&D in GDP in Germany and the United States is around 3 per cent – most Asian economies are at around one-fifth to one-tenth of that level. India's spending on R&D is

---

[29] This is a term developed by Christopher Freeman (1995) to describe the network of institutions whose activities and interactions initiate and diffuse new technologies. See also OECD (1997).

Table 5.1 *Innovation inputs and outputs in Asian economies and the United States*

	Bangladesh	China	India	Indonesia	Malaysia	Pakistan	Philippines	South Korea	Thailand	Vietnam	United States
Innovation inputs											
Global Innovation Index 2019, rank *Proxy for innovation conditions*	116	14	52	85	35	105	54	11	43	42	3
R&D expenditure (% of GDP), 2018	n/a	2.19%	0.65%	0.23%	1.44%**	0.24%***	0.16%*	4.81%	1.00%***	0.53%***	2.84%
Innovation outputs											
Number of scientific and engineering articles published, 2018	3,135	528,263	135,788	26,948	23,661	12,904	2,237	66,376	12,514	4,286	422,808

169

Table 5.1 (cont.)

	Bangladesh	China	India	Indonesia	Malaysia	Pakistan	Philippines	South Korea	Thailand	Vietnam	United States
Ratio publications/1M population	19.43	379.30	100.39	100.68	750.47	60.81	20.98	1286.19	180.24	44.87	1294.23
Number of patents granted, 2018	138	432,147	13,908	6,374	4,287	265	3,435	119,012	3,818	2,219	307,759
Ratio patents/1M population	0.86	310.29	10.28	23.81	135.97	1.25	32.21	2306.14	54.99	23.23	942.06
Unicorns, total	0	121	21	5	0	0	1	10	0	0	227
Ratio unicorns/1B population	0.00	86.88	15.53	18.68	0.00	0.00	9.38	193.77	0.00	0.00	694.85

Note: Latest available data: *2015; **2016; ***2017
Source: WIPO (2019); WIPO, Cornell University and INSEAD (2019); CB Insights (2020); World Bank (2020f)

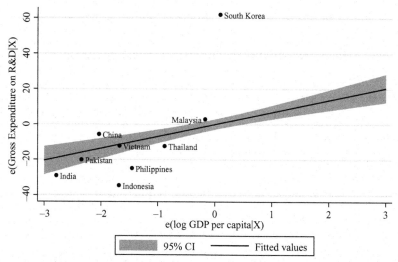

The figure shows the correlation of Gross Expenditure on R&D with real GDP per capita controlling for the size of the economy (GDP) in 2019.

Data source: Gross Expenditure on R&D- Global Innovation Index, Real GDP per capita in 2010 US$- World Development Indicators (WDI)
Number of countries: 113

FIGURE 5.1 Correlation of gross expenditure on R&D with real GDP per capita controlling for the size of the economy, 2019

at barely 0.6 per cent of GDP which is below countries such as Brazil (1.3 per cent) or Turkey (1 per cent). But there are two important exceptions. China spends around 2.2 per cent of its GDP on R&D, and South Korea spends a massive 4.6 per cent.

What does Asia look like when compared to other countries at similar levels of development once we take account of a country's size? In Figure 5.1, we undertake an exercise similar to those in Chapter 2, comparing the R&D situation in particular Asian economies against the 'rest of the world', defined by the correlation line between R&D activity and the level of development. The figure therefore shows the correlation line for the relationship across 113 countries between R&D expenditure and the log of GDP per capita, taking account also for the level of GDP. The figure not only contains a regression line but also a 95% confidence interval (CI), within which all but 5 per cent of observations are contained. The slope of the line is positive which indicates that across the world,

R&D expenditures are greater in richer countries. To get at Asia's comparative position, each major Asian economy is explicitly identified. There is only one country in the region that operates way above international norms and that is South Korea. China and Malaysia are also slightly above the 95% CI. Two countries perform more or less as expected given their level of development: Pakistan and Vietnam. But four countries – India, Indonesia, Philippines and Thailand – undertake significantly less R&D spending than would be expected given their size and level of development.

These findings are perhaps not surprising. Relatively less developed and smaller economies will face barriers to R&D as researchers and equipment are not only costly but there tend to be relatively few people in these countries with the relevant training and skills (and even they are susceptible to the lure of superior research labs and salaries overseas). While the United States educates more than 88 per cent of its population to the tertiary level, the range in Asian economies (excluding South Korea) is typically much lower, between 28 per cent (India) and 50 per cent (China). Asian levels are a bit more in line with those in economies such as Brazil and Romania (around 50 per cent). Asian countries also have relatively low proportions of tertiary students studying science and engineering (typically around 20–30 per cent). One consequence is that the research-intensive universities that provide the fuel for innovation are still mostly located in advanced economies.[30]

5.3.2 Performance in Science and Patents

Returning to Table 5.1 we can see how Asia does in terms of scientific research output, an important input to any innovation ecosystem. As a point of comparison, the United States publishes more than 400,000 articles in scientific and engineering journals per year. Most Asian countries publish a tiny fraction of that. Total publications per annum

---

[30] Some Chinese universities – mainly in Beijing and Shanghai – have recently begun to break the mould.

are less than 5 per cent of the United States total in Bangladesh, Pakistan, Thailand and Vietnam, and less than 10 per cent in Indonesia and Malaysia. However, the underlying science and engineering base is far stronger in South Korea, and especially in India and China. In fact, China actually exceeds the US level. One index that tracks high-quality research and collaborations in the natural sciences found that in 2017 Beijing was the top contributor region in terms of the authorship of papers in the eighty-two high-quality research journals, followed by New York and then Boston.[31] However, when looking at scientific output in per capita terms, research output in South Korea is greater than the United States, whilst China and India drop to around a third and a tenth, respectively.

Patents represent another way to measure the output from R&D. These are very low for most Asian countries, less than 1000 per year in Bangladesh and Pakistan and less than 5000 in four others. This compares with more than 300,000 in the United States. In line with its R&D intensity (R&D expenditure as a share of GDP), India also generates relatively few patents, especially when considered in per capita terms; only around 10 compared with almost 1000 in the United States. Again, South Korea and China are the exceptions. The latter has more patents than the United States (though far fewer per capita) while the former has almost half the US level (and rather more therefore on a per capita basis).

However, with many of these patents, the emphasis is on incremental innovation: patents of radical inventions remain rare. One measure of this comes from intellectual property rights receipts in total trade. In the Philippines and Indonesia, these are zero and reach only 0.2 per cent in China. The exception again is South Korea where the share is more than 1 per cent. However, this is still modest by US standards (5 per cent) and is also somewhat below Brazil.

Another important indicator of innovative infrastructure, especially in digital and ICT space, is the proportion of the population

---

[31] This is the Nature (2020) index.

with access to the Internet. For reference, this is around 84 per cent in the United States. In Asia, India is at 38 per cent, China at 61 per cent and it is only South Korea which is at US levels. Again, most of Asia also does not do well even compared with middle-income countries such as Brazil and Romania (60 per cent and 72 per cent, respectively).

5.3.3   Global Innovation Index

Pulling these and other indicators together in one consolidated index for national innovation is done in the WIPO Global Innovation Index. It places most Asian economies in the middle, between 40th and 80th position, although Bangladesh and Pakistan, rank far lower. As such, most Asian countries sit below transition economies such as Hungary and Romania, but in a similar position to a bunch of middle-income countries – Mexico, Brazil, Iran, Morocco and South Africa. Two Asian countries are ranked quite highly in the Index by international standards, China and South Korea. India, which ranks fifth in the world in terms of GDP, is only ranked 52nd in the Index.[32]

Figure 5.2 now uses this information and shows a correlation line for the relationship across 127 countries between the Global Innovation Index and GDP per capita.[33] As before, we have drawn in the regression line and the 95% CI and picked out each Asian economy separately. Several Asian economies, including South Korea but also China, India, Vietnam and Philippines, are significantly above the global regression line in terms of innovation. But there are also a lot of differences – or heterogeneity – across the region, with Indonesia and Bangladesh performing below what one might expect for their level of development, with Pakistan, Malaysia and Thailand tracking international norms.

---

[32] Despite a strong showing in knowledge and technology outputs (ranked 24th), India is let down by its education (ranked 102nd) and its infrastructure (ranked 60th).

[33] Note that the figure is presented in log form. We do not control in this figure for the size of the economy because the dependent variable is an index rather than a measure of expenditure which is influenced by the size of the economy.

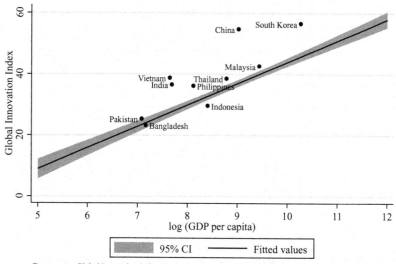

FIGURE 5.2 Correlation of the Global Innovation Index with real GDP per capita (logs), 2019

What can we conclude? While no country in Asia is yet world-leading in innovation, South Korea and China perform considerably above levels that might be expected for their level of development. India is also important for its scale, although not in per capita terms. It has also had an impact in a few key sectors, such as software development. The remaining countries mostly perform around or below what might be predicted based on their levels of development and the fact that they are in the efficiency stage. We next explore the process of technology diffusion from advanced economies and then consider local innovation. Because China, South Korea and, in some respects, India all look different, we will also explore in more depth how innovation has been evolving in these countries.

5.4 DIFFUSION OF INNOVATION

If much of the scientific invention is taking place in more technologically advanced economies, an obvious question is how countries in the efficiency-driven stage – as is the case for most Asian economies – obtain their technology? Firms have always operated technologies or produced products under license from Western firms but in recent years, one of the most common ways in which technology has been transferred has been via FDI inflows. Western MNEs create subsidiaries or joint ventures into which they place and manage their technologies and products, producing locally to meet local or global needs. They thereby transfer the technologies or products developed in their home base and train workers to operate and manage the new machines and systems.

5.4.1 Diffusion through FDI

The obvious starting point is to see how Asia has performed in terms of attracting FDI. To that end, Figures 5.3 and 5.4 summarise the situation both for the relationship between the stock of FDI and the size of the economy as measured by GDP, as well as between FDI and

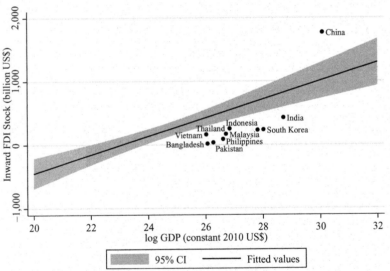

Data source: Inward FDI Stock- UNCTADstat, Real GDP in 2010 US$- World Development Indicators (WDI)
Number of countries: 173

FIGURE 5.3 The relationship between FDI stock and GDP, 2019

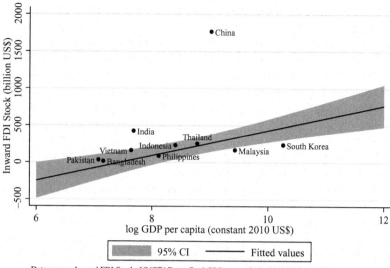

Data source: Inward FDI Stock- UNCTADstat, Real GDP per capita in 2010 US$- World Development
Indicators (WDI)
Number of countries: 173

FIGURE 5.4 The relationship between FDI stock and GDP per capita, 2019

the level of development, as measured by GDP per capita.[34] As
expected, the figures show that larger economies receive more FDI
and that FDI is greater in more developed economies. However, what
the figures also show is that most Asian economies receive significantly less FDI than could be expected given the size of their economy, although that is less true for the level of development. In terms
of size or GDP, the only exception is China, which lies significantly
above the 95% CI; all other economies lie significantly below the line.
However, once GDP per capita is considered, FDI is around the level
achieved in comparable economies, except in Malaysia and South
Korea which are below the line and India and China which are above.
These results are surprising as it could have been expected that Asia's
rapid growth would have been associated with high levels of FDI in a

---

[34] Both GDP and GDP per capita are given in logarithmic form in the figures.

mutually supportive manner, both causing and accelerating technology diffusion.

The explanation is that Asian economies have deliberately chosen to keep levels of FDI below levels consistent with their size and level of development. The most significant barriers are direct rules and regulations hindering or preventing foreign MNEs from setting up shop or forcing them to enter only through joint ventures. In India, foreign ownership stakes were restricted to limited sectors and capped at 40 per cent until the late 1990s. Subsequent liberalisation has led to an increase in FDI inflows, especially in the past five years. China, Vietnam and Indonesia have also historically imposed the requirement that foreign firms enter only through joint ventures with domestic partners. Such arrangements are obviously open to manipulation within the connections world, with domestic partners being chosen for their relationship with the government rather than their broader suitability. As such, the joint ventures requirement has been a significant policy tool deterring FDI throughout the region. Among other instruments, Indonesia and the Philippines have created Negative Investment Lists and other restrictions on foreign equity limits, investment locations, size of investors and types of licenses. Vietnam does not have a Negative Investment List, but there are sectoral restrictions and decisions about investment above a certain size are referred to the prime minister's office. Bangladesh, Pakistan and the Philippines also have numerous restrictions on FDI. For Asia as a whole, the OECD has found that it is more restrictive for FDI than other regions. Their index puts together information on the level of foreign equity permitted; screening and approval procedures as well as restrictions on foreign personnel and landownership. The index with its scale from zero (open) to 1 (closed) has an average value in Asia of 0.22 as against the non-OECD average of 0.15 and an OECD average of 0.07. A few Asian countries in fact exhibit very restrictive FDI regimes, notably the Philippines (0.40), China (0.36) and Indonesia (0.32).

Wind energy in China provides an interesting example of this phenomenon. China is a world leader in wind power development as the country tries to overcome concerns with pollution and climate change. It has more than 145 GW of wind power installed, more than all of the European Union. One might have expected considerable involvement of Vestas, one of the world's leading wind energy companies with a more than 30 per cent global market share and a manufacturer in China since 2009. However, the Chinese wind industry has been infamously difficult for foreign companies to break into and it has been difficult for any Western companies to gain market share because SOEs – the main purchasers – have historically favoured domestic manufacturers. At the same time, Chinese industrial and state procurement policies have favoured the rapid growth of domestic producers, such as Goldwind.

Even if Asia is less open to FDI than might have been expected, it is also true that the diffusion process for innovations has actually become much more varied and subtle over time. Technologies also diffuse as a result of competitive pressure that induces domestic firms to imitate foreign ones or entrepreneurs to create new firms to replicate and enhance those innovations. Some of this is driven by workers moving between firms: from foreign-owned to domestic firms, or from firms based overseas back to the home country – so-called returnee entrepreneurs and workers. In doing this, they can capitalise on their knowledge of the new technologies and combine, perhaps more effectively than within the MNE, with their understanding of the local marketplace. Examples include Kavin Bharti Mittal who worked at Google and Goldman Sachs before founding his own mobile messaging app, Hike Messenger in India in 2012; a company valued at $1.4 billion in 2016. Such processes have often been accelerated by entrepreneurs who have returned from abroad.[35] The most notable instances have been the software firms that have benefitted enormously from the return of Indians with technical education acquired

---

[35] A 2011 Report by the Kauffman Foundation estimated that more than 150,000 Indian and comparable numbers from China were returning from the US to their home countries: around half intending to start a business.

in the rich world, principally the United States. These returnees have provided both an exceptionally skilled labour force and acted as a source for domestic entrepreneurship in the sector.

5.4.2   Diffusion through MNEs

Despite these constraints, MNEs remain central to innovation. Again, the Indian software industry is a good example. This industry started in the 1980s with IBM and other MNEs choosing India as the base for low-cost business process outsourcing. In the 1990s, the country also became a destination for product development, with a large number of companies establishing R&D centres over the coming decades. Examples include the R&D centres of MNEs that still feature extensively in Indian research and patents statistics. In fact, 79 per cent of all patents granted in the United States between 1976 and 2006 to Indian-based businesses were from MNEs. In 2015, Huawei Technologies opened an R&D campus in Bangalore and, today, more than twenty global unicorns (a term for privately-held start-ups valued at over $1 billion) have R&D centres in India, a tripling since 2015.[36]

Further, although the industry was initially dominated by global players, Indian firms have now grown to become competitive and innovative multinationals in their own right. One of the first was Tata, whose Tata Consulting Services (TCS) company has now become by far the most profitable part of this venerable business group. New Indian entrepreneurial entrants have also expanded aggressively. The largest Indian IT companies, which include Infosys, HCL, Wipro and Redington India, now employ nearly a million workers. Moreover, like their international competitors, these firms are necessarily highly innovative, clearly in gradual, but also sometimes in radical, innovation. For example, TCS is working on technologies, such as natural language processing, machine and deep learning, and data marketplaces. It also focuses on innovation in the field of materials engineering, behavioural sciences and biosciences, particularly in genome

---

[36] NASSCOM (2019).

interpretation for the personalisation of medicine and developing biomarkers for early detection of diseases. Similarly, Infosys' high-tech segment, which accounts for 8 per cent of total revenues, offers cloud, consulting, engineering and IT services to various sectors including semiconductors, telecom original equipment manufacturers, consumer electronics, high-tech distributors and software vendors. There is a common path to these stories of successful and innovative domestic firms. They start out as imitators or licensees of Western technology and then gradually develop their own competences and capabilities.

A parallel story can also be told about the rise of the Indian pharmaceutical (pharma) sector. This was also once dominated by MNEs which went to India to source low-cost generic drug manufacturing. Today, the sector has over 3,000 companies and is one of the major global players in terms of both production volume and domestic consumption value, ranking third and fourteenth, respectively. Most production remains imitative and low-cost – as reflected in the fact that India accounts for 20 per cent of total pharmaceutical sales, but only 1.4 per cent of their value. Generic drugs are a key growth driver for the top Indian pharmaceuticals and represent 75–80 per cent of exports, especially to the United States. Even so, a new patent regime put in place in 2005 encouraged domestic companies to invest substantial amounts in R&D, specifically in specialty drugs and complex generics. Pharma firms are now also innovating heavily, in both gradual and radical ways. More than 65 per cent of Indian pharmaceuticals companies leverage in-house technology, while investment in R&D on average is 8 per cent of their sales, with the top companies well above that (up to 15 per cent at Lupin and Cipla). Leading Indian pharma businesses are also raising funds aggressively to fund acquisitions in domestic and foreign markets to increase their product portfolio and help with the rapid commercialisation of new drugs. Examples of research breakthroughs include Lupin, which in 2012 identified and developed arterolane maleate for malaria and India's first new chemical entity (NCE) as well as Dr Reddy's

Laboratories which has obtained 170 abbreviated new drug applications (ANDAs), over 500 drug master files (DMFs) along with 86 patents filed in the last 5 years.

Another way to acquire Western technology is to purchase Western firms in pursuit of patents, technologies and expertise. This is a rapidly growing phenomenon. For example, Chinese overseas mergers and acquisitions (M&A) reached $50 billion in 2010 and more than $150 billion in 2017. This mainly involved acquiring firms in the United States and Europe in such sectors as industrials, information technology and healthcare.[37] Among the numerous examples of emerging economy M&A into advanced countries can be included Tata's purchase of Jaguar Land Rover. Of course, some of this takeover activity is to break into new sectors, such as green energy. But it is also about upgrading manufacturing capabilities and production quality. Examples include Haier's acquisition of GE Appliances or the acquisition of German manufacturing firms, such as Putzmeister, an engineering company making concrete pumps, bought by Sany Heavy Industry in 2012; the plastics processing machinery maker KraussMaffei bought by ChemChina in 2016 and the metering company, Ista, bought by CK Infrastructure in 2017.

5.4.3 Diffusion by Domestic Firms

Although MNEs may be important in transferring and seeding new technologies, this still leaves plenty of space for local firms to diffuse new products and processes fit for local circumstances and tastes. Production processes developed for advanced economies may not always be appropriate and local processes may have to be adapted to fit the different cost environment; for example, production that is more labour-intensive and less reliant on digitisation and robots. Furthermore, there is considerable variation of taste and consumption patterns between advanced and emerging economies. One obvious example is so-called base of the pyramid product development,

---

[37] Casanova and Miroux (2018)..

whereby the relatively large unit sizes for fast-moving consumer goods, such as shampoos or soap powders, are reduced to take account of significantly lower average incomes. This has led to a wave of innovation in product size, packaging and even content. Another example is simply down to differences in taste. Daimler had always prided itself on the 'new car smell' which differentiated new from second-hand cars, and which customers in Europe and North America greatly appreciated. It discovered that this odour was unattractive in China and was reducing new car sales. As a result, they needed to innovate to eliminate the smell![38]

Innovation in Asia is also supported by a relatively open consumer mindset, probably a consequence of the relatively recent growth in real incomes and the resulting lack of a deep legacy in consumer tastes. In many Western economies, at least until the COVID-19 crisis, a significant proportion of consumers have been resistant to shopping online, or using mobile payments, even though the technology and the infrastructure for delivery has been available. However, in China, where many city dwellers are first generation, and retail memories are less deep, the rise of online shopping has been much more rapid, heavily supported by the development of the local social media/payments/B2C platform WeChat, founded by Tencent. This process is also enabled by lower levels of concern about the external use of data that exists in most advanced economies.

Particular technologies may also have a different salience in emerging economies. Take for example the case of banking. In advanced economies, banks have for generations been associated with bricks and mortar. In Asia, while there are also such banks, they tend to serve only a small minority of consumers. Most transactions are in cash. Mobile money – a form of fintech – has begun to stand this on its head because of low transaction costs, flexibility and ease of payments or transfers. This innovation, developed originally by M-Pesa in Kenya, is becoming ubiquitous in Asia, and especially in India.[39]

[38] See Estrin et al. (2020).　　[39] See Jack and Suri (2014).

India also provides numerous examples of ways in which domestic firms innovate by adapting foreign technologies to fit the local context, for example through so-called frugal innovation (*jugaad*). This refers to the widespread practice of using low-cost techniques to innovate while working within resource constraints. For example, Reliance's Jio has deepened the mobile sector in India with free offerings for low-income segments while Tata Motors introduced the world's cheapest car (costing approx. $1,500), the Tata Nano, in 2008 following the previous success of the Tata Ace. The latter was a small commercial vehicle sold at approximately $500 and designed as a 'cheap, nasty and rugged vehicle for India'. The Tata Ace generated $175 million revenue within a year and created a new market for this type of vehicle. However, the core development team was five people who worked for only twenty-two months with a total budget of $49 million. Compare this with Tesla's research budget which was in excess of $1.3 billion in each year between 2017 and 2019.[40]

5.4.4 From Technology Diffusion to Innovation

As we have seen, Asian innovation has gone far beyond adaptation – and has spawned some very successful new firms. Consider, Infosys, founded by Narayana Murthy in 1981 to provide high-quality IT services at relatively low cost. The model was premised on using talented but relatively low-paid engineers from India to serve clients in the rich world. The company quickly developed its 'Global Delivery Model', using remote teams across India to save costs. Regulatory barriers were a major hurdle in the early years. The company had to wait for a year for government permission to purchase the

---

[40] Another interesting example is the fridge – ChotuKool. This was launched in 2010 to cope with the erratic power supply in many parts of India. The product, which operates on battery, can keep foodstuffs 20 degrees below outside temperature and was designed to be suitable for rural households as well as for migrant workers and street shopkeepers. To cut costs, Godrej reduced the number of parts from 200 to 20 and sold the fridge at 50 per cent below the next cheapest model.

company's first computer and its founder had to make hundreds of visits to government offices in Delhi to acquire a variety of permits for the business. But deregulation of the sector from the early 1990s along with development of the IT infrastructure dramatically improved the situation, allowing the company to move from mainframe systems to modern technology platforms.

Infosys' innovation agenda boiled down to two core activities: improving productivity and launching new and well-integrated services. In 1999, at a time when Infosys' revenues were only about $100 million, the company established a R&D group known as the Software Engineering and Technologies Laboratories (SETLabs), which received about 1 per cent of the company budget. As Infosys' growth ambitions expanded, so did SETLabs' mission. Even if much of the company's core business was business process outsourcing (BPO), there has also been genuine innovation. In 2017 Infosys launched Infosys Nia, an artificial intelligence (AI) platform that built on the success of the company's first-generation AI platform (Infosys Mana) as well as a solution for robotic process automation, known as AssistEdge. Infosys has also invested in UNSILO, a Danish AI start-up focussed on advanced text analysis and ideaForge, an Indian start-up focussed on unmanned aerial vehicles.[41] It also acquired Kallidus, a provider of digital solutions, including mobile commerce and in-store shopping, and Brilliant Basics, a London-based product design and customer experience company.

South Korea also offers examples of successful innovation. However, predictably, these are centred on the small number of *chaebols* that account for almost 90 per cent of technology exports. For example, in the semiconductor sector almost all investments have been made by two large business groups (Samsung and LG) plus a few smaller ones (Dongbu, Hyundai and Anam). Samsung, in particular, is a massive force in innovation. It spends more than 9 per cent of its sales on R&D and holds more than 130,000 patents, 40 per cent registered in the United States, second only to IBM globally. Nearly

[41] Business Standard (2020).

three-quarters of these patents are associated with products made in South Korea.

But it is China that provides the largest number of examples of innovative technology-related firms. There are, of course, the four giants: Alibaba, Baidu, Huawei and Tencent. Alibaba is ranked fifth in the world in terms of R&D spending, behind only Facebook, Google, Microsoft and Amazon, but ahead of Apple. In terms of new patents, the company has more than 3,000 and holds strategic positions in some emerging technologies, such as blockchain where it has more than 10 per cent of all patents. Tencent's WeChat has powered the explosion in online shopping and commerce, and is currently developing new technologies in AI. Baidu is the dominant Chinese search engine and spent more than 17 per cent of its revenue on R&D in 2019. Finally, Huawei, in addition to manufacturing global consumer electronic products such as mobile phones, has established itself as the leader in the provision of 5G infrastructure generating a global revenue of more than $120 billion. The company has spent $90 billion over the past decade on R&D, amounting to between 10 per cent and 15 per cent of its revenue each year. Indeed, of its approximately 150,000 workers, some 96,000 are employed for R&D purposes.

Despite this picture of innovation in some industries and regions,[42] such as Shenzhen, it should be remembered that the bulk of Chinese industry continues to operate using relatively simple components or final products with well-established technologies. Indeed, the dynamism of the four Chinese tech giants and the ecosystem around them contrasts strikingly with the continued existence of hundreds of thousands of state-owned firms and many private firms for whom innovation is barely relevant.

5.4.5 Entrepreneurship in Asia

We have argued that innovation, and especially radical innovation, often relies on a vibrant entrepreneurial ecosystem. So, what can we

---

[42] See Williamson and Yin (2014); Yip and McKern (2016); and Appelbaum et al. (2018).

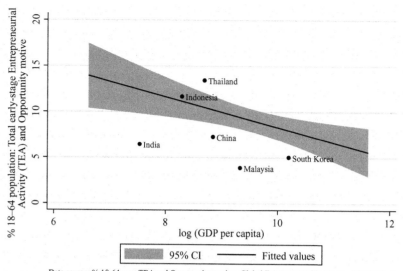

FIGURE 5.5 Opportunity entrepreneurship in Asia, 2019

say about the extent of entrepreneurialism in Asia? And how does it compare with other regions? Clearly, what is of interest from an innovation perspective is not the necessity entrepreneurship that is endemic in Asia and elsewhere in other emerging economies. That form of entrepreneurship is driven by a shortage of jobs and inadequacy of income, an issue we discuss in the next chapter. We are concerned instead with entrepreneurship which propels the economy forward, so-called opportunity entrepreneurship.[43] Figure 5.5 uses data from the *Global Entrepreneurship Monitor* database that measures the incidence of opportunity entrepreneurship across many countries. The figure contains a regression line across the sixty-four economies for which data are available to show the global relationship between opportunity entrepreneurship and GDP per capita. We see a very slight downward relationship, with the line declining from

---

[43] See for example Estrin et al. (2019).

around 12 per cent share of opportunity entrepreneurs in relatively poor economies to around 5 per cent in higher income ones.

As in previous figures, the available Asian economies are also highlighted. Interestingly, most Asian countries including China, South Korea and India fall below the regression line; there is less opportunity entrepreneurship in these countries than one might expect for their level of development. However, Indonesia and especially Thailand are above the global norm. Contrary to public perception, the evidence therefore suggests that Asia is not especially entrepreneurial, at least when measured by the number of opportunity entrepreneurs in the labour force. Yet, this only tells part of the story.

In advanced economies, entrepreneurs usually need to innovate products or processes themselves but in Asia, as in other emerging economies, entrepreneurs have mostly concentrated on the domestic market, drawing on innovations and business models developed elsewhere and applied to the specifics of the country. For instance, Jack Ma and Alibaba did not invent electronic business-to-business and business-to-consumer marketplaces but drew on the experience of US companies and diffused know-how developed elsewhere, helped by the so-called Great Firewall. Even so, local context was also critical, especially the need for Chinese small- and middle-sized enterprises (SMEs) to trade with one another, as the economy opened up in preparation for WTO membership. In India, examples include RedBus, which introduced online bus ticket selling in the country, Ola (the Indian Uber), Paytm (Indian PayPal) and Zomato (Indian Yelp). India has also developed a growing private entrepreneurial ecosystem in the fintech sector, which mainly focuses on mobile wallets and digital payments, with a growing universe of start-ups like Paytm, MobiKwik, Citrus and PayU. These have all been exploiting the rapid increase in the use of smartphones, internet connectivity and online shopping to integrate payment processing into web applications.

More recently, Asian entrepreneurship in China, India and South Korea has begun to change character. For example, in China there has been the emergence of supportive ecosystems for innovation. As in Silicon Valley, these try to pull together policies aimed at skills, clustering and open innovation. The broad idea behind this has been to move towards a more decentralised model of innovation, where new companies of all sizes can be spawned, and where the focus is increasingly on radical innovation and the development of new products rather than simply improved processes. The particular areas include the broad domain of green technology, whether of solar panels or wind turbines, along with battery storage for deployment with solar panels, as well as EVs. To put this in context, by 2016 China was already the world's largest electric vehicle market, accounting for half of the 700,000 vehicles sold worldwide that year. At the same time, China's e-commerce market accounts for around 40 per cent of the value of global transactions, up from less than 1 per cent a decade ago. And China now has 121 unicorns valued at $284 billion which is only outnumbered by the United States. All this has been supported by a fast-expanding venture capital industry, with funds in excess of $100 billion – the second largest VC industry in the world. This has fuelled waves of domestic IPOs.[44] However, while some of the major US VC players were heavily involved at the beginning, changing policy has meant that the industry is increasingly domestic, led by Alibaba and Tencent, but also with significant financial backing by the state.

In a similar vein, there has also recently been a major acceleration of the Indian start-up ecosystem, leading to a 12–15 per cent increase in number of start-ups (2014–19). As in China, this has partly been driven by an explosion in consumers going online. At the same time, there has been a large inflow of entrepreneurial and technology talent. Nearly 40 per cent of India's start-up founders come from leading engineering colleges and business schools, such as the Indian Institute of Technology and Management. Finally, large capital inflows from MNEs and global investors, as well as a rapidly

[44] Lin (2017).

expanding VC industry, have further catalysed an entrepreneurial spirit. Some business groups have established corporate-level R&D labs. One such example is the Mahindra Group with its $250 million 'Mahindra Research Valley'. Others have opted for betting on small businesses through corporate venturing, such as the Godrej Group, which has invested $50 million in businesses innovating in the group's current sectors of operation. In addition, many small businesses are being founded as spin-offs by former MNE employees. MNEs are also starting to become more involved in Indian VC, either as limited partners or as direct investors. Qualcomm, Intel and others have had corporate VC offices in India for a long time and have, more recently, been joined by Google, Cisco and others. This explains why the principal entrepreneurial hubs (Bangalore, Delhi, Mumbai, Chennai, Hyderabad and Pune) are often co-located around clusters of foreign MNEs.

Needless to say, South Korea also has a nascent entrepreneurial ecosystem. This has developed largely because of public policy combined with a highly educated labour force. However, most of this entrepreneurial activity is concentrated spatially – in Seoul – as well as in the IT and e-commerce sectors (over half of all start-ups). Most start-ups still rely on domestic markets and only a few have succeeded in generating sales outside South Korea. Even so, this upswing in entrepreneurialism has begun to yield fruit. South Korea already has ten unicorns scattered across six sectors, with four in B2C e-commerce as well as in fintech, health, travel and retail sectors.

We can conclude that in parts of Asia and especially in some sectors and cities, we are beginning to see innovation rather than technology transfer, and some of that innovation is even radical. We are also observing – at least in a few locations – the establishment of an innovation ecosystem, including venture capitalists with deep pockets.

5.5   THE ROLE OF THE STATE AND INDUSTRIAL POLICY

In China, Vietnam and South Korea, the state has traditionally led the way in trying to innovate and in the transfer of the fruits of this work to the private sector. In India, the state also took on a prominent role after Independence in 1947, although this enthusiasm has waned over the last thirty years. But whatever the country context, in the numerous exchanges on the viability and desirability of industrial policy, arguments for state-led or -facilitated innovation have mostly assumed that such investment would be made available at low or no cost to private, profit-seeking companies.[45] But this assumption, as we have argued, cannot be taken for granted in the Asian context. Indeed, the wider experience with the state as innovator (including in the former Soviet Union) suggests that while the state can sometimes successfully lead on focussed and priority projects, its high level of centralisation detracts from its ability to deliver R&D across a broad front, particularly when priorities compete. Further, those competing priorities tend to be linked to the interests and respective bargaining powers of incumbent groups. This can seriously complicate and potentially distort the setting, let alone attainment, of R&D goals.

In a Soviet manner, China under Mao single-mindedly pursued state-led innovation but over the past three decades, it has shifted its focus. The country has moved towards a state-guided, although for the most part privately implemented, innovation system which aims to achieve a more complex, higher productivity and value-added economy on the basis of long-term industrial plans. For example, in the discussions around the next Five-Year Plan taking place in 2020, it became clear that a key target of policy would be to replace, as a result of deteriorating trade relations with the United States, Chinese reliance on US microprocessors along with continuing emphasis on AI, robotics and green energy.

Recent advances in innovation – some of which we have already mentioned – are therefore not the result of recent or spontaneous breakthroughs but are the outcome of sustained and very costly public investments into science, research facilities and education. These public investments have also triggered private investment.

---

[45] As, for example, by Mazzucato (2015).

The Chinese Communist Party has viewed technological advance as a priority in order to enhance the country's power and prestige, to reduce strategic dependence on Western technology and as a basis for long-run sustainable development as demographics turn against continued extensive growth. Thus, whether innovative activities are undertaken directly by the state itself; via state-owned firms; through large private firms and business groups; or via new entrepreneurial organisations, the state has played a leading role in guiding the availability of resources and the direction of travel. This has proceeded in the context of a political stability that has provided the benefits of continuity and a long-term policy perspective: a phenomenon of considerable value in the complex process from invention to innovation. But it would be misleading to think that the application of large resource firepower has been the main, or only, attribute.

5.5.1   Protecting Domestic Industry

Crucially, as we have seen, the state has also pursued protectionism and ensured that foreign players have been kept out of key sectors that it wishes to develop. The most notable example is the so-called Great Firewall that has kept Google, Facebook and Twitter out of China, facilitating the rise of Chinese equivalents like Tencent's WeChat. Equally effective, if less draconian, has been the support – financial, customers and through regulations, tariffs and taxes – that has provided major advantages for domestic over foreign players. Thus, eBay found it very difficult to compete against Taobao and Alibaba in China, in no small part because of the advantages offered by the national and local governments to the domestic players. eBay had to shut its Chinese operations in 2010. Similarly, Uber found itself competing with a profitable and – critically – a very well-connected local competitor, Didi Chuxing. Uber ended up selling its Chinese business to Didi in 2016.

China's willingness to exercise regulatory authority on a highly discretionary basis also ensures that those domestic companies who have garnered support, and sometimes protection, from the state do not themselves embark on strategies or actions that contradict the aims of the government and the Communist Party. As a measure of the sort of control that this Faustian pact implies and in a stark departure from data privacy rules that exist in most countries, the Chinese state has the right to have access to highly intrusive information about these companies' clients. That access has, in turn, been linked to a whole system of so-called social credits that have been used by the government as a way of influencing, if not controlling, citizens' behaviour.

What also stands out in China is that in the newer dispensation, although the selected private firms are clearly different from state-owned companies, they are still highly politically connected and have to be willing to be guided by state priorities and norms of behaviour. Even limited challenges to those prerogatives can be met with actions that remind private companies of who calls the shots. The last-minute, aborted flotation of Ant Financial discussed previously is a stark case in point. And so is the way that Didi has been mauled in 2021 for choosing a foreign listing.

In short, each of these new generation companies has, in a variety of ways, very strong links to government, not only because of funding and regulations but through protection from competition, domestic or foreign. For these companies, the benefits of compliance are also clearly very substantial. Baidu's search engine, for example, obviously benefits massively from the prohibition of Google, Facebook or other non-Chinese search tools. Similarly, freed from foreign competition, Tencent dominates gaming, social media and is a close second to Alibaba in mobile payments in China. Presently, around over 1.1 billion people use Weixin and WeChat on a monthly basis. Alibaba also leads in e-commerce and has over 750 million mobile monthly users. Tencent and Alibaba's duopoly accounts for a 62 per cent market share of China's cloud infrastructure market.

Across the Himalayas, the Indian state has also historically been an active user of industrial policy, but it has played much less of a role in directing and supporting innovation than China. Instead, the Indian state has opted for a role managing the rules of the game and to a lesser extent, influencing the paths of the main players. In the case of the nascent IT sector, which we discussed earlier in the chapter, the state did not regulate heavily but, rather, created tax incentives for technology imports and exports as well as establishing limits on foreign ownership. These measures allowed Indian firms, entrepreneurs and software developers to emerge, develop networks and be positioned to exploit the opportunities for software services when MNEs searched for outsourcing companies. As such, the sector grew without large state investment and, arguably, outside of the purview of most state policy, especially the dense web of regulations faced by firms in more traditional sectors.[46]

The evolution of India's pharmaceutical industry has also been defined by a number of key public policies. In the 1970s, the state enacted a patent act which allowed the pharmaceutical firms to obtain their own patents. As there were no restrictions on intellectual property rights, many Indian pharma firms started reverse engineering for non-infringing processes. These dynamics made the Indian market much less attractive to the MNEs, enabling Indian companies to carve a niche in both the Indian and global markets based on their expertise in drug manufacture at low cost. Subsequently, policy has pushed Indian businesses to do more R&D through foreign investments and collaboration with MNEs.

South Korea has long been held up as an example of how industrial policy can be used to nurture and grow both sectors and businesses.[47] Among these, the rise of South Korean high-tech companies from the 1960s onwards was to a large extent the consequence of a sustained long-term government strategy. The latter supported capital-intensive industries and contributed to the strengthening of technological potential, the creation of research institutions and, of course, to the development of the system of higher education. The state offered incentive systems for private in-house R&D and

---

[46] See Commander (2005).   [47] See, inter alia, Amsden (1992); Rodrik (2009).

underwrote loans to fund projects being executed by the main business groups. This, combined with a highly competitive market environment, kicked off a (technological) race among those business groups.

The rise of the South Korean mobile phone sector to global prominence is an interesting example of the interplay between the state and *chaebols* in the innovation process. In 2007, the market share of Samsung in South Korean smartphones was already 50 per cent, largely because mobile phone makers benefitted from protectionism in the pre-iPhone era with the state blocking the entrance of foreign competition and supporting the development of local smartphones. Samsung was chosen by the state to be the national champion in this market, with associated privileges and the requirement to make the state a major stakeholder. Currently, the switch towards greater openness to foreign competition along with other public policy decisions, such as deregulating the WIPI (Wireless Internet Platform for Interoperability) indicates that the state may be willing to play a less prominent role in defining the direction of this particular sector.

However, the South Korean state also intervenes strongly to support an entrepreneurial ecosystem providing finance and tax exemptions to new industries and start-ups. In 2018, 13 per cent of all South Korean small businesses benefitted from the government's support policy, mainly funding. Indeed, 98 per cent of funds raised by small businesses actually come from the state rather than capital markets. This generosity with funding is doubtless one factor behind the 50 per cent increase in the number of South Korean start-ups between 2011 and 2017. Even so, the hegemony of the *chaebols* has not yet been significantly challenged by new entrants.

Other policy instruments aimed at sponsoring innovation have been quite widely adopted through Asia. The most notable are the attempts at increasing the technological spillovers from such FDI as is permitted. China led the way through the widespread use of special economic zones (SEZs). These comprise science and industry parks and/or bonded zones, normally accompanied by low taxes and financial incentives for FDI, as well as local entrepreneurship, with public financing of infrastructure such as Internet, ports and airports. The basic aim is for the government to create and subsidise clusters of knowledge dissemination, innovation and entrepreneurship with the model being loosely that of Silicon Valley. Shenzhen is the most-cited example as China's tech centre and location for the headquarters of its major technology companies such as Huawei and Tencent. In all, China has six SEZs which are responsible for around 45 per cent of FDI and 60 per cent of exports. This success in managing MNE entry and ensuring successful diffusion of technology to domestic firms has spawned many imitators, including in India, Malaysia, Thailand, Vietnam and, potentially, Indonesia. However, China's initial success with this policy, which rests in part on a high degree of state direction of the economy, has so far been hard to replicate elsewhere.

Let us summarise our findings and argument so far. Asian economies are not undertaking high levels of R&D nor are they leaders in global innovation, although a few are beginning to play a significant role in key sectors on the global stage. Furthermore, many Asian governments have been effective, given their level of development, in their use of longer-term policies aimed at creating conditions conducive to innovation. In some places, such as India, these policies have been rather ad hoc and piecemeal; in others such as China and South Korea, the state has been far more directive, selecting key sectors for development and channelling resources via tax breaks, subsidies and soft loans in their direction, mostly to preferred business groups. While MNEs have been welcomed and have played a major role in technology transfer, their acceptance has been qualified. Barriers to foreign companies are common across Asia not least because governments have tried to favour local companies. Finally, although there is evidence of rising opportunity entrepreneurship – something that can be expected to contribute to innovation – the Asian countries still mostly punch below their weight. We now turn to looking explicitly at the role that the connections world has played and is likely to play with respect to innovation.

5.6 INNOVATION AND THE CONNECTIONS WORLD

5.6.1 Business Groups and Innovation

The limits on FDI and resulting protection of domestic firms against foreign ones that we have discussed earlier in this chapter are not a random outcome. The closely intertwined political and business elites that comprise the connections world have mutual interests in restricting or tailoring entry by foreign MNEs in ways that do not subvert or threaten the powerful market positions of domestic incumbents. Business groups thrive on the rents generated for domestic firms by the restrictions on imported products and technologies. At the same time, political elites can win popularity from pursuing policies of economic nationalism which restrict the role of foreigners in a country's economic development. In China, such populism has been increasingly allied in recent years to a fairly explicit policy of autarchy for key sectors, thereby reducing dependence on Western technology and supply chains.

The restriction of FDI would not necessarily have substantial deleterious effects on technology diffusion and innovation if the opportunities thereby created stimulated innovation by domestic firms. To some extent, there is evidence that this is the case. Business groups such as Tata and Samsung are leaders in domestic, and in some areas, global innovation. Some of the new business groups in China, such as Tencent or Alibaba, or in India, Wipro or Infosys, are clearly technological powerhouses. Business groups can, it seems, be very effective at innovation particularly when working closely with their governments, as in South Korea and China, and when huge financial and human resources are applied to the problems at hand.[48] When government is not there to support and drive innovation, as in India, then the traditional, incumbent business groups mostly do not innovate massively but in those sectors or activities where incumbents are not all-powerful, there can be significant amounts of innovative activity.

---

[48] For discussions of cases, see Yip and McKern (2016), Appelbaum et al. (2018).

That business groups can, and do, innovate is hardly surprising because of their superior access to resources, including finance and managerial capabilities, relative to non-affiliated companies. In fact, studies have found that typically business groups do innovate more than other domestic players.[49] This is in part because, as we discussed in Chapter 4, business groups can help to resolve market failures in countries where institutions are weak, Not only do they create internal markets for capital, skilled labour and management but their deep pockets from having market power at home and across a wide range of sectors gives them the ability to fund the high levels of R&D spending that is observed in companies such as Huawei, Samsung and Tata. As we have seen, the role of some business groups as 'national champions' for key strategic sectors, such as Samsung in South Korea and Huawei for 5G technology, also opens the door to huge amounts of direct subsidy and soft loans provided by state-owned banks.[50]

But this apparently rosy picture belies other facets that are altogether less positive for Asian innovation. In particular, it is not obvious that business groups have the incentives, but also the flexibility of organisational structure to generate, support and absorb the new knowledge that is at the heart of innovation. We already noted that incumbent firms everywhere struggle with innovation and especially radical innovation because the organisation is inherently resistant to the introduction of products and processes that threaten its current business model. Such conservative organisational tendencies, already visible in Western firms such as Xerox or Kodak, are likely to be more pronounced in the large and complex structures of contemporary Asian business groups, where managers struggle for the attention of the owning dynasty. Expressed differently, this raises the question of whether business groups have the organisational capacity to recognise and assimilate new external information.[51]

---

[49] See, for example, Belenzon and Berkovitz (2010); Castellacci (2015). For evidence on India, see Castellacci (2015); Komera et al. (2018).
[50] See Nolan (2001); Lane (2021).   [51] Cohen and Levinthal (1990).

Such organisational capabilities, often referred to as absorptive capacity, capture the competences of an organisation and tend to be cumulative.

Indeed, the picture is altogether more problematic for business groups. Perhaps the most important factor concerns the underlying objectives of the organisation, and its position within the connections world. As we have seen, business groups are best understood as having been constructed to exploit the weaknesses in institutions that typify many developing economies. They bypass inefficient market structures and allocate resources – notably capital and management – internally. But ultimately these resources are allocated across affiliates to satisfy the desires of the owning family, for example to accumulate private wealth and to obtain personal prestige and political power. Rent-seeking and the exploitation of market power is never far from the list of objectives. Yet, innovation is a disruptive activity. It threatens to undermine established monopolies and accumulated market and non-market power. As such, innovation may be anathema to many of the vested interests within the business group, who would see their profits, as well as their personal positions, brought into question by fundamental changes to the way business is done. Indeed, business groups are often constructed to funnel resources and profits from the periphery – the diverse bunch of corporate affiliates – to the core which is the family holding. This does not inherently make them averse to new technology or knowledge, but there is a predisposition against innovation if the new technology undermines carefully built-up positions of political and economic power.

This inherent conservatism of many business groups is likely to be exacerbated by their governance arrangements. It will be remembered that the business groups' ownership arrangements are opaque, and shareholdings do not necessarily reflect underlying control because of pyramid-like structures, which gives the dominant family control rights in excess of their ownership rights. Although most ordinary shareholders will be primarily motivated by profits and will therefore be keen to undertake and exploit innovation that increases profitability in business groups, they are minority shareholders and hence often lack influence. In some instances, the majority shareholders – families and their public and private corporate vehicles – may share these objectives. But when business groups exist primarily as vehicles for transferring profits within the organisation and ultimately to the benefit of their (family) owners, this will not be consistent with maximizing the profits of each individual affiliate. This is because the owner's plan for the affiliate may not be innovation, even if promising, but instead the transfer of profits to other affiliates. The main point is that Asian business groups tend to be structured to maintain the political and economic dominance of the group itself, rather than the efficiency or performance of the individual affiliates. Motivations and associated behaviours of this sort are generally inconsistent with creating a culture of innovation.

In short, although business groups commonly have significant resources, they do not necessarily provide the incentives or structures to exploit them for innovation. Further, with concentrated ownership, there will tend to be little appetite for risk and little leeway given for failure by the owners. This will discourage experimentation and innovation. The consequence will not be an organisation open to learning and the assimilation of knowledge and technology from outside. Nor is the arcane business group organisational structure well-designed to apply new knowledge and information internally. Moreover, management will have been selected for its networking talents and ability to nurture and grow rents. With that in mind, it is perhaps hardly surprising that while numerous patents now emanate from Chinese and South Korean firms, the bulk of them are not international patents and most concern process rather than product innovation: in other words, gradual rather than radical innovation.

Even so, there are two important areas in which innovation might be strongly in the interest of the group owners. The first is in technologically advanced sectors where the pace of change is rapid and the second is in the face of overseas competition. Thus, companies such as Alibaba and Tencent in China, and Infosys or Ranbaxy in India, do not rely only on connections, supportive regulations or protection to bolster their profits. They need to innovate massively to stay ahead of the competition, domestic and global, and – in these instances – the business group format, if anything, helps them to do that. At the same time, the relatively few business groups that earn their profits on the international market, for example Samsung or Hyundai from South Korea, once again have little choice but to invest huge sums in innovation in order to compete in global markets.

However, these do not represent the bulk of Asian business groups who congregate in more traditional sectors – agribusiness, chemicals, natural resources, construction, transport and logistics – where the pace of radical technological change is slower and the benefits of close relationships with the government greater. It is problematic for governments to regulate sectors undergoing rapid technological change because it is hard to discern the final form of the technology. By contrast, the full panoply of government–business relations, including subsidies, soft loans and procurement, can be provided in traditional sectors in return for assistance in meeting national targets, social goals or creating jobs. It is in these sectors that the deleterious impact on innovation will be most marked. Moreover, even when business groups move into new sectors, they often retain their opaque structures and governance with portfolios that continue to comprise activities where the connections world is still the main currency. That is certainly the case for the two most successful Indian business groups of recent times – Reliance and the Adani Group. In China, the fundamental tension that is emerging is between the state's promotion of innovation in private businesses and its apparently growing desire to control or influence those businesses.

5.6.2 Suppressing Competition: Business Groups and Entrepreneurship

We saw in Chapter 4 that a key characteristic of an economy dominated by business groups is that levels of market concentration and overall concentration are very high. There is a longstanding view in economics that highly concentrated markets will generate a much slower rate of innovation. This is partly because, as we have seen previously, the incentives to innovate are much less sharp when a firm already has significant market power. We have argued that these arguments are more pressing when the dominant firms are affiliates of a business group. But market dominance brings another problem which is highly damaging to the trial-and-error process also central to innovation and especially radical innovation. This is due to the high barriers to entry by new ventures. As we have seen, the process of innovation often relies on experimentation by new organisations, many of which will actually fail in their attempts to launch a new product or process. This process of creative destruction is at the heart of innovation but is severely limited by high entry barriers, because these makes it harder for new firms to establish themselves and grow.

There are numerous examples of such problems. Business groups, with their privileged access to both financial resources and government loans and subsidies, can ensure that new entrants are starved of funds. Business groups also offer secure and well-paid employment to managers and skilled workers. Such arrangements are rarely on offer in new ventures. Their absence in environments where there is little social insurance will accentuate the risky nature of choosing to work in a new venture. Further, business groups will have built their market power via relationships with suppliers and with retail outlets, both of which will be hard for entrants to match, even if the incumbents do not actively exploit their relationship and connections to choke off any potential competition. However, in fact exploiting their market power and contacts to choke off entry appears common. Newspapers in South Korea are filled with stories of business groups preventing access to new entrants. The same abuse of market dominance applies to other critical facets for the creation of new ventures, such logistics, marketing, IT and business services.

Yet, all of this ignores perhaps the most serious problem for new entrants in the connections world. These are the barriers to entry put in place by the government, often at the behest of business groups.

The political connections of business group owners can be called upon to introduce new rules and regulations to limit new firm entry. They range from minor obstructions, such as zoning or fire regulations, to outright prohibitions against new entrants, both domestic and especially from overseas. In short, there is an uneven playing field between incumbents and entrants at the best of times. Against business groups with their resources, financial muscle and connections in both the business and political world, the odds are tilted even more strongly against new ventures. India provides an important example of these issues. As we have seen, while a minority of business groups have become engaged in new sectors, most have remained firmly entrenched within the confines of their traditional spheres of business. For the most part, therefore, the Indian corporate world comprises business groups sustaining themselves and supporting the families that own them and their clients. Although considerable innovation has occurred in some sectors, these are usually outside the industries and locations that are subject to the stranglehold of the connections world. Moreover, the firms leading these breakthroughs have often been new ones. The question remains whether they will remain highly innovative as the benefits of incumbency and the advantages of the connections world increasingly become available to them. It is perhaps the huge size of the Indian economy which has helped innovation through the emergence of new firms in new sectors, based in new regions. The economy has enough 'space' for newcomers to grow away from the dead-hand of the connections world. But in the more traditional sectors and organisations, innovation has gained little traction.

Suppression of competition is also significant in China where government policy drives an innovation strategy based on national champions. Huawei may have many of the characteristics of an open-source innovative company, with research labs in India and the United States as well as China, and internal incentives and structures designed to increase absorptive capacity. But at the same time, its national champion status and associated and favourable access to state funding and procurement also represents a major barrier to entry for possible competitors, domestic or foreign. More generally, as we saw in Chapter 4, China has recently seen a considerable concentration of economic power and consolidation of wealth in a small number of hands. Although the pliable connected companies that are willing to play the game may innovate – this framework leaves little space for new entrants.

Turning to South Korea, the large business groups are certainly undertaking substantial innovation, partly under the pressure of international competition.[52] However, the question is whether there would be more innovation, or at a lower cost, in a more competitive and open market structure. The fact that innovation is concentrated in a small number of huge firms means that the South Korean economy is not very diversified in terms of innovation: focusing the bulk of innovative activity in the *chaebols* increases the risk that judgements about the future path of technological development may prove to be incorrect.

South Korea also provides numerous examples of how the dominance of business groups crowds out innovation by new entrants. The majority of smaller South Korean businesses are incorporated into the *chaebols'* supply chains; hence, exclusion from the supply chain represents business extinction. For example, in the electronic and car manufacturing industry, businesses supplying to Samsung and KIA operate as first or second tiers of assembling and are required to comply with the company's requirements. This hampers the capacity of small firms to innovate because the supply chains function in a top-down way. Moreover, there is a tradition in South Korea of incumbents crowding out new entrants as the *chaebols* often establish their own affiliates to compete and undercut start-ups. As a result, South

---

[52] Taiwan was also one of the four so-called Asian tigers of the 1990s. Although its growth was technologically intensive, it was not necessarily innovative: a characteristic that has continued to the present day. Taiwan is also now facing serious challenges, notably, lack of scale due to its small domestic market as well as its complex and antagonistic relationship with its giant neighbour, China

Korea still mostly lacks a dynamic entrepreneurial ecosystem. The small business sector is made up mainly of very small companies and only a few of them grow even to medium size. Small businesses have focussed on enhancing competitiveness through cost reductions, have lower levels of productivity – only 30–40 per cent of the productivity of large businesses – and do not offer well-paying, long-term employment opportunities.

5.7 CONCLUSIONS

How well positioned are the Asian economies to base their growth on innovation? In this chapter we have argued that contrary to some commonly held views, most of them are not very well positioned for such a transition. But this is not necessarily a problematic conclusion. Indeed, most Asian countries lack the institutional, political and human capital framework to move beyond being recipients of new technologies largely developed elsewhere. But they are not necessarily passive recipients; technology diffusion often involves significant adaptation of products and processes to local conditions. This point is hammered home by Indonesia's four, current unicorns. All of them have succeeded by extending advanced economy business models to Indonesia's particular environment. Thus, *Go-Jek* is an Uber-style online transport service; *Traveloka*, an Expedia-type booking platform; and *BukaLapak* and *Tokopedia* are eBay-like online marketplaces. It may be too early for most Asian economies to become world-class innovators but there remain numerous profitable opportunities from efficiency-driven growth.

However, somewhat different questions about the future of innovation are raised in India, China and South Korea. Each is currently nudging at the boundaries of innovation-driven growth. South Korea seems to have alighted on its own unique path to innovation-driven growth. The high levels of education and skills have been the fundamental drivers, along with the increasing willingness of the government to pass the baton for economic development to the business sector. South Korea is consequently a very innovative place, although the emphasis remains on gradual rather than radical innovation. However, this innovation is typically done in a highly concentrated way that still knits together a small number of large business groups and the state. Beneath these large trees, little else has flourished so far. Even so, the selection of priorities and sectors has proven very effective and incentives for conservatism and entrenchment have mostly been overcome, not least because of the discipline exerted by participating in competitive export markets. However, the main risks for this national innovation strategy lie in the fact that almost everything is concentrated in business groups and in their capacity to anticipate broader changes and to adapt effectively. For example, to address the likely worldwide replacement of internal combustion by electric engines for automobiles in the next decades, the *chaebols* producing vehicles have largely eschewed the global technological alliances among suppliers and battery manufacturers that have emerged in the United States and Europe, opting to rely on their own technological capabilities and supply chains. This is a highly ambitious but also very risky strategy in the face of a potentially disruptive new technology. As the large, incumbent business groups still exert a stranglehold over much of the economy, the obvious danger is that in the future they may prove ineffective or unsuccessful at innovation, and especially radical innovation, but are yet able to stifle it from new entrants.

Turning to India, we have seen the emergence of some global players among the business groups as well as an entrepreneurial ecosystem in a few large cities. However, the Indian government has not been effective in building a widespread education and skill base while, for the most part, the private sector has not developed a strong scientific base capable of generating widespread inventions of its own. Thus, while there is a great deal of innovation in India, much of it is derived from overseas and transmitted through MNEs or brought back by returning migrants. Moreover, the bulk of innovation takes the form of adapting inventions from elsewhere to suit the needs of the huge domestic market. Unlike in China, the Indian state does not lead

invention nor guide economic development in a coherent or systematic way. Indeed, despite reforms, industrial policies still contribute
significantly to restraining entry and limiting market competition. As
we have seen, innovation rests largely in the hands of incumbent
firms, usually business groups, often enjoying the benefits of protection from foreign competition, along with regulatory controls helping
them to establish domestic market dominance. For some of these
business groups, interested in efficiency and profitability, the concentration of key resources, notably finance and skilled labour, enables
them to innovate. For others, located in more traditional sectors
and entrenched in the body politic of particular regions, entering
higher-risk activities involving innovation is not attractive. Either
way, the advantage of incumbency stacks the scales against entry by
new firms.

The issues are very different in China. The country is placing
immense resources behind its target to become a global technology
leader. The process it is following may be both very costly and inefficient, involving the targeting of financial and other resources to
sectors and national champions selected in particular sectors but at
least so far, it has been quite effective. The Chinese approach to
innovation has several strands. The first is a Soviet-style approach
with the state guiding innovation activity in public sector laboratories
and state-owned companies. It then supports diffusion through its
control of the financial system, through its procurement policies
and through its network for incentive and control that works through
the Chinese Communist Party. These activities are also strengthened
by systematic efforts to obtain Western technologies and know-how
through FDI and joint ventures, and through acquisition of, or from,
leading Western technology firms. A related strand of policy acts to
persuade highly trained scientists and engineers from the United
States and Europe to return to China, to work in companies such as
Tencent and Alibaba or to create new entrepreneurial firms of their
own, often financed by state-influenced VC funds. In these ways, the
Chinese state drives innovation in what it regards as strategic sectors.

However, despite its systematic character and the vast financial resources being made available, the approach has only worked some of the time. Thus, there is no doubt that China has successfully created widespread digital business-to-consumer and business-to-business markets. In some ways, visitors to China find that it is more advanced than many Western economies in digital applications in retail, services and finance. This partly reflects the advantages of being able to leapfrog legacy technologies and delivery mechanisms such as bank branch networks or traditional logistical systems. Moreover, as noted previously, because high levels of consumerism are relatively recent, Chinese consumers are less resistant to change. In terms of strategic industrial policy, China has also successfully developed green energy (solar and wind), battery and robotic companies. However, China's agility in adopting new technologies and consumption patterns is likely to slow as the current digitally based systems become entrenched. Moreover, the industrial strategy has had failures as well as successes, for example in automobiles and airplanes. In both key sectors, despite decades of effort centred around creating national champions, the gains have been very limited. The global aircraft industry remains a duopoly between Boeing and Airbus, despite both companies having joint venture subsidiaries in China. Similarly, even the huge Chinese car market remains dominated by Western manufacturers. Chinese efforts to build electric cars attractive to world consumers, in competition with Tesla, have also so far met with limited success.

The innovation landscape in China is not just populated by large, state-owned or state-influenced companies. The second prong in the innovation strategy involves actively developing clusters for private sector innovation. At first glance, these resemble the advanced economy model with large numbers of new ventures, often emerging from university research laboratories, supported by an entrepreneurial ecosystem, including private VC and networks of business support services. However, the relationship between these new private entrepreneurial firms and the more traditional state-owned firm–business group nexus often remains unclear and certainly, not transparent. As mentioned previously, the space in which these different innovators operate may be quite distinct, with private ventures effectively left to develop areas in which incumbents and their networks are not present. In this case, decentralised forms of innovation may in the future tend to be restricted to the early stage of the product cycle, with the new organisations that are created being gradually absorbed into the connections world that dominates the remainder of the economy. Moreover, China's increasingly authoritarian turn is in direct contradiction to the type of environment that nurtures scientists and creative people and sponsors innovation. This will surely make it harder to motivate inventors and entrepreneurs who will be increasingly wary of the political risks from challenging entrenched businesses and political interests. At the same time, Chinese emigrees working in laboratories in the United States and Europe may find the climate less attractive, despite the promise of unlimited resources, and therefore prefer to apply their education and talents outside China.

An alternative view is that this entrepreneurial energy will presage a gradual evolution of the entire Chinese economy to a more open, less politicised and competitive market structure. But for this to happen will require a changing political climate and a policy-driven dismantling of the huge pockets of influence and rent-taking – the essential attributes of the connections world – that exist.

6 Employment in the Connections World

Governments worldwide have long trumpeted increases to national income as the metric of success, not least in Asia. Indeed, as we saw in Chapter 2, most of the talk about Asia's economic miracle has been cast in terms of income growth. That single focus has been perhaps most relentless in both the Chinese and Vietnamese variants of capitalism for – as Branko Milanović (2019) has observed – nothing much else validates their systems. Yet, behind this almost fetishistic and overt emphasis on growth lies an ever-present, dominant and unsettling preoccupation for politicians: the constant need to create jobs and to meet the demand for employment by citizens. This is because political legitimacy in the case of the autocracies depends on creating employment and preventing unemployment, perhaps more so than raising incomes. This is particularly relevant in China where household spending has been deliberately held down as a matter of policy. In the case of the democracies, voting behaviour tends to be influenced by the state of voters' employment and income prospects, especially in urban areas. In short, whilst growth delivers a supportive public narrative with many desirable consequences – not least the reduction of poverty – its employment consequences are mostly what matter to those in power. This cuts across political systems. The seemingly limited priority given in public to employment by politicians seriously belies its significance.

Yet, creating jobs, let alone productive ones, is far from being a mechanical correlate of growth in income, especially in developing economies. Moreover, the ways in which the economy is configured, along with the ownership and governance of companies, can have profound – and often long-lasting – consequences for employment creation. In this chapter, we argue specifically that the way the economy is configured reflects the powerful imprint of the connections world which has major consequences for both the scale, and type, of employment that is created. This is because the connections world across Asia not only supports the prevalence of business groups and SOEs, thereby impeding the emergence of new companies and activities, but also because it limits the capacity of the formal economy to grow much beyond those companies that are connected. What we mean by the formal economy is that part of the economy where firms and labour are taxed and subject to regulation, especially concerning employment rules and standards such as working hours, conditions and redundancy. By contrast, the informal economy largely falls outside the tax and regulatory net. Despite growth in the formal economy being a much-vaunted feature of public policy over many decades, this has mostly remained in word rather than deed. In practice and in part because of the connections world, Asian governments have found it hard to achieve their ambitious employment goals and this has led not only to high levels of informal employment and to perpetuation of the gulf between formal and informal employment. Workers face wide disparities not just in their conditions of work but also in their respective earnings and productivity. One consequence of this has been to boost income inequality.

In a nutshell, the privileged world of formality for the most part remains populated by relatively small numbers of business groups and their constituent parts, along with public sector companies, government employees and a limited number of other foreign- and domestically-owned companies. However, the great mass of domestic firms remains small and relatively unproductive offering precarious work while still accounting for very large shares of these countries' labour forces. The perpetuation of this duality can be seen as one product of the connections world and its wider modus vivendi, notably the way in which entry and growth of new companies is held back while small- and middle-sized firms are forced into ancillary roles in the supply chains of large competitors or as sources of cheap outsourced labour. Moreover, as we shall see, what is also so striking in much of Asia is the absence of organic growth through small firms expanding over time. One consequence is that medium-sized firms form a relatively small layer. Large firms are also far less present than in most of the richer economies. These gaps limit potentially powerful sources of economic dynamism.

6.2   THE EMPLOYMENT CHALLENGE

Before considering in greater detail how Asia's employment is organised, let us start by getting a sense of the employment challenge it has faced and will continue to face. Asia currently accounts for around 60 per cent of the world's labour force. Between 2000 and 2020, just to maintain a stable employment share – the ratio of those employed relative to the labour force – the whole of Asia has had to create over 2 million jobs a month. To appreciate these magnitudes, even with its persistently low birth rate, China has had nearly 9 million students graduating from college and entering the labour market in 2020 alone. Looking forward, in the coming twenty years up to 2040, despite lower projected rates of labour force growth and ageing populations in almost every country, to sustain a stable employment share, Asia will still have to create around a million jobs a month.

Further, despite strong productivity growth over recent decades,[1] labour productivity in Asia, as we saw in Chapter 2, is on average still a small fraction of the levels achieved in the rich world. GDP per worker in East and Southeast Asia is only around 15 per cent and South Asia less than 10 per cent of that in the advanced economies. The reasons for this gap embrace a number of factors including market structure and the extent of rivalry, but also the ways in which

---

[1] Between 1997 and 2017, labour productivity throughout Asia increased on average by around 4.5 per cent annually, with East Asia having nearly 5.5 per cent annual growth.

economic activity is organised in Asia and, especially, the continuing and massive presence of the informal economy.

The scale of Asia's future employment challenges will be made more difficult when considering the implications of the introduction of new technology and associated ways of working. We noted in the previous chapter, for example, the way in which China is, among other initiatives, trying to achieve global leadership in artificial intelligence (AI), along with the accelerated use of robots in the workplace. These can be expected to have an impact not only on the volume of labour but – perhaps most importantly – the type of labour that is required. Recent experiences of technological change – notably the introduction of ICT – have tended to shift demand towards more skilled types of labour rather than simply displace workers. But there is no denying that the options facing unskilled labour are likely to deteriorate.[2] In Asia, the employment and wage prospects of the very large numbers of unskilled and semi-skilled workers are already poor, so any further move to disadvantage them in the labour market will be worrisome – with implications for inequality – and potentially destabilising. We return to these challenges in the concluding chapter.

6.3 ASIA'S DUAL ECONOMY

There have been dramatic changes in the composition of output in Asia. This has also affected employment. Figure 6.1 shows that everywhere there has been a substantial decline in the share of agriculture in total employment since the early 1990s, along with an accompanying rapid urbanisation. Although, industry has grown – notably since 2000 in Vietnam – what is quite striking is that industry's share has mostly stabilised or even retreated in the last couple of decades. In contrast, most Asian economies have seen a sizeable increase in services. This has been particularly prominent in South Korea, Malaysia and, more recently, in China. But within this broad process

---

[2] What Goldin and Katz (2007) memorably referred to as the race between education and technology.

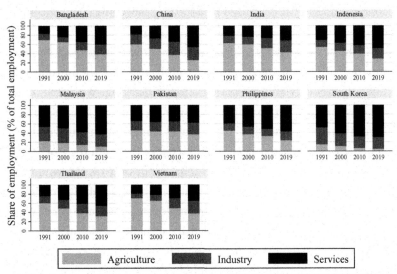

Data source: Employment in agriculture, industry and services (% of total employment) (modeled ILO estimate)-
World Development Indicators (WDI)

FIGURE 6.1   Share of employment by sector, 1991–2019

of structural change, the conditions under which people work has continued to follow a divide that has been long present; namely, the coexistence of two broad sectors of the economy outside of (but also within) agriculture which are, as we have already mentioned, commonly bracketed as formal and informal.

This coexistence is sometimes rolled up in the term, a dual economy. What this means is that an economy is comprised of one part – the formal – subject to labour regulation and rules regarding wage setting, working conditions and taxation, such as payroll taxes. Such workers are normally found in business groups, foreign invested companies and state-owned enterprises, particularly the larger ones, as well as the public administration. The other part – the informal – includes family members working without pay in a family concern, the great bulk of the self-employed and, often, even a significant share of paid employees. Self-employment often elides with specific forms of entrepreneurship, especially, what we termed in the previous chapter, necessity entrepreneurship.[3] Indeed, in Asia, the ILO estimates that over 90 per cent of entrepreneurs are informal and most of these are driven by simple necessity. Although there is some variation across countries, many informal sector workers and entrepreneurs have relatively little education – over 50 per cent of informal workers in Asia have no, or only primary, education[4] – and work through very small firms oriented towards local markets. Moreover, they tend to receive very low wages and perform at low levels of productivity. The reasons for this may lie not just in the workers' lack of training and skills but also their inability to secure affordable finance. Informal, unregistered firms or individuals commonly lack access to capital markets and banks and hence have to rely on cash and/or creditors, some of whom may be family. Whatever the source, the cost of capital tends to be high, sometimes cripplingly so. Nevertheless, the informal economy in Asia does include numerous small family businesses that often work as lower tiers of larger, formal companies. But whatever the exact configuration, the common reality remains that informality almost always implies operating way within potential, and it is precisely this gap that has long motivated economists, policy advisers and governments to promote the translation of informal into formal activity. That this has largely failed to happen is damaging but, as we shall see, hardly surprising in the context of the connections world. The incentives for expanding formal work have remained far too limited.

The dual nature of the economy can be visualised in a simple diagram (Figure 6.2). In the upper parts of the figure, there are two differently sized shapes. The larger represents the informal sector; the smaller the formal one. Some of the main attributes of the two sectors are noted. Probably the most significant distinction relates to the low levels of job security, access to benefits, wages and productivity in the

---

[3] These are entrepreneurs who start businesses out of necessity – to survive. Examples include food market and street sellers, bicycle and motorbike taxis as well as small-scale handicrafts.
[4] ILO (2018).

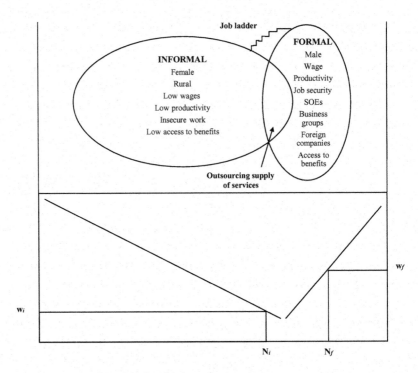

FIGURE 6.2 Formal and informal sector

informal compared to the formal sector. If these two parts of the economy are pretty much self-contained, it can be said that segmentation exists. However, it can be seen that there are actually two points of connection. One consists of a staircase or ladder by which workers from the informal economy move up to jobs in the formal sector. That this occurs is not in doubt. But what is also clear is that the numbers that go up this staircase are rather limited. The second point of intersection is more direct. That is where formal sector firms draw on informal workers – sometimes through explicit outsourcing – to perform particular tasks or functions. This intersection varies significantly across countries and can be one way in which the formal sector benefits from the relatively cheap and flexible labour in the large informal economy.

The lower part of the figure also provides an economy-wide representation. On the horizontal axis there is employment and on the two vertical lines there are wages. In this simple representation, informal employment is significantly larger than formal employment while the wage level in the latter is significantly higher than in the former. The amount of open unemployment which – as is common in most developing and emerging economies – is relatively small. Chronic underemployment, associated with the size of the informal economy, is a more dominant feature of these economies. Underemployment also means that labour productivity stays low because workers are unable to function at their full potential with, as a corollary, mostly low – if not miserable – earnings.

6.3.1 Dimensions of the Dual Economy

What do the Asian economies actually look like in terms of the formal/informal divide? Most attempts at measuring the size of these different components have attempted to compute their respective shares of GDP. For obvious reasons, this has not been a straightforward task. Even so, it appears that at the start of the 1990s, in South Asia, the share of the informal economy was around a third and in Southeast Asia around two-fifths. In China and Vietnam, the contribution was 18 per cent and 22 per cent, respectively.[5] The richest country – South Korea – was thought to have an informal sector that accounted for just under 30 per cent of GDP at that time.

Fast forward to 2015 and these shares of GDP had declined significantly everywhere, mostly by between 20 and 35 percentage points. For example, China had its informal economy contributing only 12 per cent of GDP, as against 18 per cent in India or under 20 per cent in South Korea.[6] These paths do broadly suggest that as income has risen, the output share of informality has duly fallen. What they

---

[5] Historically, planned or communist economies have tried to suppress informality, whether in the former Soviet Union or in China and Vietnam. One consequence was the need to create more formal jobs, not least through a bloated public sector. In more recent times, both China and Vietnam have done little to discourage informality .

[6] Medina and Schneider (2018).

also imply, of course, is that the informal economy's declining share of output has a lot to do with the very low value added or productivity of informal firms, let alone their pervasive inability to grow. Hence, economic growth has for the most part not come from the informal economy but from the formal one. In other words, informal firms may be very numerous, but they mostly operate at very low levels of capacity and efficiency. Consequently, while measuring their contribution to GDP is important, it would be hardly surprising if that share was far smaller than the informal economy's share of employment or, for that matter, the total number of firms in the economy.

The evidence suggests that it is indeed the case that the share of the informal sector in employment is much higher than for output. Further, in contrast to what we saw concerning shares of GDP, the share of informal sector employment appears not to have been falling but actually rising. Quite how much informality has risen is not so easy to compute as statisticians and others have not measured employment in the informal economy over time very well or consistently. Nevertheless, when pulling together what is available, in around 1990 in South Asia between 65 per cent and 75 per cent of employment was thought to be informal. This was also the case for the Philippines and Indonesia, while in Thailand the share was thought to be around 50 per cent.[7] For other countries, there is an insufficiency of evidence.

Turning to the present, Table 6.1 shows that the informal economy is very much a feature of all these economies. The formal economy – with a major role played by business groups and the public sector – still mostly accounts for a relatively small share of total employment outside of agriculture. In South Asia, the share of informal employment ranges between 71 per cent and 82 per cent when excluding agriculture, whilst in Southeast Asia it ranges between 34 per cent and 80 per cent. In China and Vietnam, informal employment accounts for 54 per cent and 72 per cent of total employment,

---

[7] Charmes (2012).

Table 6.1 *Formal and informal share of employment, 2018–2019*

Region & country	Income classification	Informal share (%)	Formal share (%)
South Asia			
Bangladesh	LM	82	18
India	LM	78	12
Pakistan	LM	71	29
Southeast Asia			
Indonesia	UM	80	20
Malaysia	UM	34	66
Philippines	LM	56	44
Thailand	UM	55	45
Vietnam	LM	72	28
East Asia			
China	UM	54	46
South Korea	H	29	71
Comparators			
Egypt	LM	50	50
Morocco	LM	76	24
Ghana	LM	63	37
Argentina	UM	47	53
Brazil	UM	43	57
Russia	UM	36	64
France	H	9	91
Japan	H	16	84
USA	H	18	82

Key: LM=Lower Middle: UM=Upper Middle: H= High Income
Sources: ILO, IMF and authors' calculations

respectively.[8] Put differently, when thinking about jobs, the shiny, modern world of Indian software companies' campuses or the thriving

---

[8] The shares do not change dramatically when including agriculture – where most work is classed as informal. In South Asia, for example, the range increases to 82–89 per cent.

commercial and industrial hubs created in some of China's special economic zones, whilst deeply impressive, are very far from the norm.

How does this compare with countries at similar levels of income? Table 6.1 also shows that for those classed as Lower Middle Income by the World Bank, the informal sector employment share is mostly significantly higher than for other comparable countries outside of Asia. This is true for most of the upper middle-income countries as well. For example, China has a 10–15 percentage points higher share than Argentina, Brazil or Russia. Even in the richest economy – South Korea – informal employment still comprises around 30 per cent of total employment. To put this in context, Japan or the US's informal sector accounts for 16–18 per cent of total employment. What is evident is that even if Asia has managed to create new formal sector jobs – which it undoubtedly has in reasonable magnitudes – it has struggled to increase the formal sector's share of total employment. Informality, with its attributes of low wages and productivity, remains pervasive.

6.4   WHY IS ASIAN INFORMALITY SO ENDURING?

There are several possible and related responses to this obvious question. One is that regulations and taxation have blunted the incentives for informal firms to choose to become formal: the costs simply continue to outweigh the benefits. Barriers to access to finance or infrastructure – even something as basic as access to electric power – may consign firms to informality. Back in the late 1980s, the Peruvian economist – Hernando de Soto – highlighted the way in which lack of title to land and other assets not only contributed to poverty but also to ensuring the perpetuation of informality.[9] These are factors that are also relevant in contemporary Asia.

Considering the costs of entry that bureaucratic hurdles and stipulations may impose in Asia, the average time taken to register

[9] For example, in De Soto (1989).

a small business ranges between 6 days in Thailand to 33 days in the Philippines. In India and China, the number of days is 18 and 9, respectively.[10] In other words, there are surely more procedures, more days and more cost in Asia to starting a business compared with the situation in most advanced economies. Even simple details such as registering a property or paying taxes may be cumbrous and off-putting. The time taken in all parts of Asia to register a property appears to be between three and five times that in the advanced economies.

And it is also not just about starting up. As – if not more – important, will be the costs and benefits of functioning as a formal sector business. How attractive it is to create formal jobs will clearly be affected by the costs attached to both hiring and firing employees. Consider the case of labour legislation in Indonesia. Under the country's 2003 Labour Law, employers are prohibited from implementing large-scale redundancies. Even furloughs require initiation by workers rather than employers. Redundancies can only be put through with the agreement of the Industrial Relations Court or through a mutual settlement process with workers. In reality, the former is hardly feasible, so mutual settlements are the only way and these inevitably imply severance payments. Such payments are widespread in Asia, in part because of the absence of effective unemployment insurance systems. To put this in context, in Japan, the United States or Singapore, employers are under no legal requirement to pay any severance at all. In Hong Kong, that requirement is also very low – around 1.4 weeks of salary on average. But elsewhere in Asia, matters are quite different. On average, an employer has to pay over 26 weeks of salary when parting with a formal sector worker. In India, the average payment is around 11.5 weeks, in China just over 23 weeks, in Thailand it approaches 32 while in Indonesia it amounts to nearly

---

[10] These numbers are drawn from the World Bank's annual *Doing Business* (2020b) publication.

58 weeks! These costs form part of the reason for why creating jobs in the formal economy may not be attractive.

While such severance payments can help give workers a measure of greater job and income security, if they also make the ease of entry and exit of firms more difficult, they will have wider consequences of an efficiency nature. Lower entry can suppress the arrival of new firms who, in many instances, may be more dynamic and productive. Similarly, making it more difficult for firms to exit will also impede the elimination of weak or failing firms. Indeed, a wide range of studies across many economies, including in Asia, have found that more entry and exit is generally associated with greater allocative efficiency. These changes represent essential features behind the growth of productivity that Joseph Schumpeter memorably termed, creative destruction.[11] This is because both processes act in a rather Darwinian manner while also supporting the introduction of new products and services.

And this matters also for employment. Obviously, when entry is impeded, most creation and destruction of jobs will occur in existing companies. If exit is hard to organise then destruction of jobs may be lower but by the same token the creation of jobs is also likely to be lower. This is not a marginal phenomenon either as, even with these restrictive arrangements, the creation of jobs in new entrants has been quite significant. In Indonesia between 1990 and 2009, new entrants and start-ups were actually the largest contributors to employment growth.[12] The reality is that larger firms in Asia do not create large number of jobs, even if the ones they do create are mostly 'good' as measured by wages and productivity and, commonly, job security.

These explanations for why the formal economy has struggled to grow its share of employment in Asia over a protracted period of time clearly play some part. Yet, it is hard to imagine that factors such as the cost of setting up a business or firing workers are the only, or necessarily, the main explanatory factors. For instance, informal firms

---

[11] Schumpeter (1942); Bartelsmann et al. (2004).   [12] Javorcik et al. (2012).

are mostly fragile and unproductive as they are often run by poorly educated people with limited skills and networks. Although the efficiency of informal work is susceptible to improvement through experience and educational attainments have mostly been rising, this has made limited inroads in boosting skills and productivity. As a consequence, they will struggle to survive in the formal economy. When there is a yawning gap in capacities and productivity, a corollary is that it really does not matter very much if barriers to entry to the formal sector – such as the cost of registering an activity – are high or low. There is this more profound gap that separates the two parts of the economy and results in the sort of segmentation that we discussed earlier. However, educational enrolments and attainments have generally risen significantly and the efficiency of informal work is susceptible to improvement through experience, something that is not always limited by poor education.

A complementary line of argument is that the boundaries between formal and informal sectors and their persistence may well be strongly affected by the interests of the players and, specifically, by the way in which formal firms elect to operate. Some may, for instance, prefer to have a large informal economy from which to draw cheap and unregulated services and labour when required, as suggested by the intersection of informal and formal sectors in Figure 6.2. Certainly, there is evidence that large companies will outsource work or activities to informal players. In Chapter 5, we noted the way in which South Korea's *chaebols* do indeed draw on smaller companies as part of their supply chains. However, the most obvious and widespread example concerns the garments industry. In the Indian subcontinent, it is common for textile concerns to draw on home-based workers and subcontractors in informal entities. Home-based workers are particularly poorly remunerated, receiving wages between 50 per cent and 90 per cent below the minimum wage.[13] Chains of subcontractors commonly resort to low pay, poor working

---

[13] Kara (2019).

conditions and few rights for workers. A shocking example was the fire in the Tazreen Fashions factory in Dhaka, Bangladesh in 2012 which killed over 100 people. It turned out that the firm was completing orders for the huge US company, Walmart, while functioning as part of a chain of outsourcers.[14] But whilst for some industries this may amount to a useful and cheap strategy, the amount of outsourced work will naturally be held back by the limitations of the informal sector, including those of a managerial, financial and skill nature. Moreover, when informal firms are constrained to be small, the transaction costs of subcontracting will be high; often prohibitively so. As such, this form of self-interest is not really a very convincing explanation for the latter's widespread persistence.

All these various explanations for the perpetuation of informality have aspects that undoubtedly ring true but taken together fail to tell a wholly convincing story. This brings us back ineluctably to the central themes of this book. As we have seen, a critical feature in the functioning of the connections world is the way in which incumbents strive so hard to maintain market power and entrench themselves. At the same time, the public sector companies that also benefit from their direct connections to power have mostly been established on the basis of monopoly or, at the least, a substantial amount of market power. The rub of it is that the business groups and allied companies that are so prominent in the formal parts of these economies are quite content to limit entry as best they can. And in this, they have been remarkably successful. But before we examine how the connections world acts to conserve the sharp bipolarity between formal and informal economies, it will be helpful to get a better sense of the company landscape and, in particular, the size of firms and their distribution across these various economies.

6.5 THE SIZE OF FIRMS

In Asia, as elsewhere, the vast majority of firms – commonly over 95 per cent – are micro and small firms. This host of small companies

---

[14] Walmart had apparently removed Tazreen from its list of approved factories.

and individual operators also accounts for a very substantial share of employment, over 80 per cent in Bangladesh and as much as 94 per cent in Indonesia. Although some small firms are in the formal sector, many – often the majority – are not. In India, only half of all small and micro firms are formal. In the Philippines, micro and small firms also comprise just over half of formal employment but their share in total employment (which includes the informal economy) is far higher. However, being precise about this is complicated as the data are scarce, often unreliable and sometimes organised in ways that defy easy comparison.

Even so, it is clear the landscape of firms in the Asian economies has some common features. If thought of as a pyramid, the base is composed of huge numbers of these small, if not tiny, units as in India with its 63 million micro and small firms employing over 110 million people.[15] Further – with the exception of South Korea, the great majority of these small firms are in the informal economy. Above them, sit a far thinner layer of medium-sized companies – the great majority of which are formal – that mostly account for 5–20 per cent of formal employment. At the top of the pyramid sit a small group of large companies,[16] whose contributions to employment and output are often far from trivial. In Southeast Asia, for example, large companies account for over two-fifths of GDP![17] As to employment, in the Philippines, large companies account for 18 per cent of formal employment, although that shrinks to no more than 7.5 per cent of total employment. In Indonesia, those shares are 20 per cent and 3 per cent, respectively. Even in the richest economy – South Korea – with its powerful *chaebols*, large firms account for no more than 20 per cent of formal employment.[18] China appears still to be a bit different in that large companies – often state-owned ones – employ over half of

---

[15] In the Indian context, micro firms are considered to be those with less than ten employees and small firms less than fifty.

[16] Different statistical agencies use different cut-offs but, for the most part, large companies are those with >200 employees.

[17] BAPPENAS (2018).   [18] OECD (2018b).

the formal sector. Statistical and political reticence about the informal economy makes it harder to compute this as a share of total employment.

How does this compare with the situation in advanced economies? Evidence compiled by the OECD shows that in the rich world, micro and small firms also comprise the largest share of firms – between 90 per cent and 95 per cent – while medium-sized firms average around 5 per cent and large firms <1 per cent.[19] This is not where the difference lies. It is when considering employment that the picture changes drastically. In North America and Europe, large- and medium-sized firms combined account for nearly 60 per cent of total employment, with the greater part in large firms. There are variations, for sure. In Italy, for example, micro and small firms have over 60 per cent of employment and large firms just under a quarter. But in France, the United Kingdom and the United States, large firms are very prominent with 47–58 per cent of employment. Whatever the variation within these advanced economies, it is very clear that the situation is in stark contrast to that in Asia where medium and large companies occupy a far less prominent space in employment space. Does this matter?

The answer is unequivocally that it does. This is because large and medium firms tend to have some significant advantages, not just in terms of their performance but also in the ways that they are managed and treat their employees. In addition, their relative absence also tells us something about the way in which companies can develop over time and the space for growth. Evidence from a group of middle-income economies, including some in Asia, shows that large companies are more likely to be more productive, to innovate and export more, whilst also offering better working conditions, higher job security and higher remuneration for their employees.

---

[19] The OECD classifies large firms as those with over 250 employees and medium-sized firms between 50 and 249.

Once the characteristics of those employees are taken into account, the wage premium of large firm employees in parts of Asia – many of which are employed by business groups – as well as in other middle-income economies averages 15 per cent. This wage premium rises to 25–50 per cent for those firms that have over 300 employees.[20] In India, larger firms (>200 employees) pay on average five times more than firms employing between 5 and 50 workers and roughly double medium-sized entities. Large differences are also present in the Philippines. In Thailand and Malaysia, the differences range between 15 per cent and 25 per cent. For South Korea, the gap is also pronounced – the average wage in a small company is approximately half that in a large one and the productivity gap is even larger. Further, medium-sized firms also pay between 17 per cent and 25 per cent more than the small ones.

Differences in wages are even more accentuated when comparing firms in the formal economy with those in informality. For example, in Bangladesh being in formal employment in either industry or services results in earnings between 56 per cent and 63 per cent larger than for those in the same sectors but informal. In India, formal sector wages are roughly double those in informal work. Such large differences are also the case in China, with the additional twist that the large number of migrant workers in the cities suffer from particularly low wages,[21] as many lack a work permit (the *hukou* system which limits eligibility to live and work in major cities) and hence are locked out of formal sector employment. In Vietnam, formal sector earnings have been at least a third larger than informal ones. These pervasive differences in earnings are not simply a reflection of 'bad' behaviour by employers or the lack of labour rights for informal workers (although these are by no means irrelevant) but primarily a reflection of profound differences in productivity.[22]

---

[20] World Bank (2020e).   [21] Liang et al. (2016); Park et al. (2012).
[22] ADB (2012) and Rahman et al. (2018).

Then there is the additional matter of what has been described as the missing middle, the relative absence of mid-sized firms. An example is Indonesia where formal sector firms employing between 20 and 100 people comprise a very small share of firms in manufacturing.[23] Although by no means universal, many other Asian economies also have middles that are under-represented. This suggests that firms find it difficult to grow or, in some instances, preclude growing because of the regulatory costs that would be an accompaniment to crossing an employment threshold. Indeed, the available evidence shows unequivocally that as firm size increases the regulatory burden tends to become greater. One recent estimate argues that the regulatory protection of incumbents in middle-income countries, including in Asia, is 40–60 per cent higher than in the advanced economies.[24]

6.5.1 Changes in Firm Size over Time

This brings us naturally to the crucial matter of how companies grow – or do not grow – over time. In most advanced economies, the evidence shows that a significant number of companies can work their way up through the ranks, so to speak. The paths to growth are various but can be steep. How steep depends a great deal on policy. For example, in the OECD, young firms may account for about 20 per cent of employment but in recent decades they have created almost half of new jobs. Beneath this statistic lies a great deal of variation. For example, in Europe, after five years between a quarter and a half of start-ups have failed, between a third and three-quarters still have fewer than ten employees and roughly one in eight have over ten employees.[25] But a small share of entrepreneurial start-ups has taken off and they have been spectacularly important in creating jobs – for example between a quarter (Netherlands) and over a half of all new jobs (France and the United Kingdom).[26]

---

[23] Javorcik et al. (2012).   [24] World Bank (2020d).   [25] Calvino et al. (2015).
[26] OECD, DynEmp (2016).

To get these rates of growth and job creation depends, of course, on supportive policies and encouragement, something that has not always been forthcoming. In addition, the evidence suggests that not only can small firms evolve into medium-sized ones, but that medium-sized firms themselves are able to grow and, in a limited number of cases, evolve into large entities. Again, there is a lot of variation by country, sector and across time. That this can be important is probably best illustrated by Germany where medium-sized firms (50–249 persons employed) represent a small fraction of businesses (>2 per cent) but account for around 20 per cent of jobs and value added and are, on average, more productive than large firms. More generally, established medium-sized enterprises that innovate and grow tend to play an important role in aiding smaller firms to improve and participate in supply chains.[27] In sum, the broad picture from the advanced economies – despite large variations across them – is that firm growth across the different layers of size can and does occur and in so doing can make a major contribution to growth in employment and productivity.

The picture is rather different in the Asian economies. The relative absence of large firms attests to a multitude of factors, whether limits on foreign investment, excessive regulation or lack of competitive opportunities due to the entrenched power of incumbents. But it also attests to the limits of organic growth from below. An absent middle also holds pointers about the ways in which firms – both incumbents and start-ups – evolve over time. What is clear is that most small firms struggle to establish themselves and grow. This is not very different from the advanced economies. What is different is that the proportion of firms that do succeed in growing is both smaller and their rate of growth generally lower. Contrast the fact that thirty-five years after being founded, small firms in the United States had employment and productivity roughly ten times higher than at the outset, yet small firms in India declined by a quarter over a similar

[27] OECD (2018c).

period, even if their productivity rose somewhat.[28] Over a larger sample of developing countries, other research has found that barely one in ten of small firms grow to medium size and a tiny fraction (one in 100) are able to become large.[29]

In short, in Asia, the size distribution of firms – notably the massive preponderance of small and informal firms – testifies to the fact that very few are able to establish themselves and grow. The result is a landscape very different from that found in richer economies. What we see is a small number of large and productive companies – mostly part of business groups and the connections world as well as foreign-owned companies – a limited middle and a huge host of small entities, many of which are informal. Only a minority – mostly a small minority – of employees work in large private companies, some of which are not only very profitable but also dynamic. The state-owned enterprises in which many Asian governments vested so much have singularly failed to perform effectively, even if, particularly in China, they still account for a significant number of formal sector jobs. Among the consequences are that the shackles of small size, along with informality, also remain those of low productivity and earnings. This corpus of underperforming firms and individuals acts as a deadweight on the broader economy.

6.6   THE CONNECTIONS WORLD AND EMPLOYMENT

In Chapter 4, we argued that connected entities often leverage their advantages to impede the entry of challengers, while the very importance of connections also raises the costs of participation by any possible challengers. Companies that lack the appropriate ties struggle to achieve the permits and access to resources – financial or otherwise – that can allow them to compete effectively. And through this chain of barriers – small and large – the resulting distribution of firms and employment that we have just described tends to perpetuate

---

[28] World Bank (2012).   [29] World Bank (2020e).

Table 6.2 *Shares of total formal employment by country for top ten and top twenty-five companies ranked by revenues*

Country	Top 10	Top 25
Bangladesh	1.5	1.8
China	1.1	1.6
India	0.7	4.3
Indonesia	3.1	4.2
Pakistan	0.4	0.7
Philippines	0.6	1.6
Vietnam	1.1	2.6
Comparators		
Germany	6.0	11.0
Japan	2.4	4.5
USA	2.5	4.0

Source: Orbis and authors' calculations

the profound distinctions between formal and informal, let alone the distinction between small and large firms.

To get a broad sense of how business groups and other very large companies contribute to creating jobs, let us start by calculating their shares of total formal sector employment. Although, as we will see, this is an indicator that has clear limitations, it is nevertheless a revealing one when set in the context of the massive jobs challenge that faces these economies. Table 6.2 shows something very striking. For both the top ten and top twenty-five companies (as measured by revenues) in each country, their cumulated share of formal employment is very small. (If we had calculated these shares relative to total employment, they would have been even smaller, of course.) These lists include private and some state-owned companies. Take the case of the Philippines. The number of people employed by the top ten companies amounts to only 0.6 per cent of total formal employment. Expanding coverage to the top twenty-five companies does not make a huge difference, they account for no more than 1.6 per cent of formal employment. In Bangladesh, these shares are broadly comparable at 1.5 per cent and 1.8 per cent, respectively. In neighbouring India, the share is 0.7 per cent for the top ten while for the top twenty-five companies, it reaches 4.3 per cent. In China, the share for the top ten is 1.1 per cent and for the top twenty-five, 1.6 per cent. For Indonesia, the top ten companies account for 3.1 per cent of total formal employment, rising to 4.2 per cent for the top twenty-five. In the latter case, roughly three-quarters are in private companies, almost all of which are business groups and a quarter in state-owned companies. Even in South Korea, the most advanced economy in the region, the country's massive *chaebols* may account for over 40 per cent of GDP but they still employ less than 20 per cent of the formal private sector.

Put simply, these numbers show that even when some of the largest companies – ranked by revenues – employ very substantial numbers of people, they nevertheless still account for rather small shares of formal – let alone total – employment. Table 6.2 also shows that when doing the same exercise for several major advanced economies, the shares are mostly higher, particularly in Germany where the top ten companies account for over 6 per cent of employment and the top twenty-five around 11 per cent. Although the difference in the shares of total employment accounted for by the top ten or twenty-five companies between advanced and Asian economies is not that huge, in the former, the entirety of large companies accounts for very substantial shares of total employment, something that is self-evidently not true in Asia.

Before moving on, we should make an important caveat about simply looking at the employment shares of these largest companies. Their direct employment obviously grossly understates their broader consequences for jobs. That is because almost invariably – at least in manufacturing – such companies have major supply and production chains that link up with other companies. For example, Boeing in the United States not only draws on a wide range of suppliers for parts for its aircraft and other products but it also outsources major components of production, such as airframes, to other, often large, companies, such as Spirit AeroSystems. Such linkages obviously also exist in Asia.

6.6.1 Asia's Public Sector

Does the picture change when bringing in the wider public sector, including government? This is obviously a very relevant question given that so many Asian countries have at various stages put large resources into creating state-owned enterprises, as well as expanding the size of the state more generally. With respect to SOEs, the common picture is actually one of relative decline over time. In China – where SOEs have been the most numerous – total employment in SOEs was over thirty-five times larger than that of the private sector in the early 1980s. In 2006, the two were approximately equal but by 2017 SOEs accounted for only 5–16 per cent of total employment, although they still accounted for around 23–28 per cent of GDP.[30] In the past decade, Chinese SOE profits in most years have been negative in aggregate and, even when positive, always significantly lower than for private companies. In India, where there are still over 200 SOEs controlled by the federal government and over 1000 by state or regional governments, a substantial share of the latter is effectively moribund, while the remainder have also performed poorly. In the case of the centrally-owned SOEs, although in aggregate they are profitable, around a third are chronic loss makers and those that are not have far lower returns on capital than comparable private companies.[31] In short, even if the idea that SOEs would provide a leading edge for the economy has proven to be a chimera, SOEs still do employ significant numbers of people. However, their generally weak – often failing – performance has meant that the main policy challenge has been about managing their decline. Quite commonly,

---

[30] Zhang (2019). Chinese official data suggest that 18 per cent of industrial employment in 2017 was in SOEs.
[31] Chhibber (2018).

this has boiled down to trying to safeguard jobs for incumbent employees and management. Nowhere is this more visible than in the Northeast of China where clusters of declining heavy industries continue to employ very large numbers of people.

Despite the waning of SOEs in most Asian economies, the broader public sector is still often a major employer and always accounts for a significant part of the formal economy. For example, in both India and Pakistan, total public sector employment (including SOEs) makes up over 30 per cent of total formal employment and around 15 per cent in Bangladesh.[32] In the Philippines and other Southeast Asian economies, the share is in the range of 20–25 per cent.[33] Where the vestiges of socialism have been better maintained, the public sector's share is also still very high. Despite a fairly vertiginous drop in the share of public sector employees (including SOEs) in China since 1980, by 2017 they still amounted to between 25 per cent and 50 per cent of formal employment (and between 10 per cent and 22 per cent of total employment).[34] In Vietnam, the public sector share is also high at around 40 per cent of formal employment with three-quarters of that being government employment.

Although it is pretty much impossible to estimate with accuracy what is the optimal size for the public sector, there are several reasons for why its scale has wider significance. The first is normally related to the adverse impact large public sector size can have on economic performance. But this argument is far from clear-cut as much depends on the actual composition of the public sector and how much of spending on that public sector detracts or subtracts from the private sector, a phenomenon that economists have termed, crowding out. There is a second strand of argument which concerns the way in which the public sector reflects the aims and responses of politicians. This is obviously of direct relevance for us as the public

---

[32] World Bank (2018d).   [33] OECD and ADB (2019).
[34] Zhang (2019). The upper estimate (c50 per cent) depends on allocating a large residual to SOEs.

sector is an important component of the wider connections world. With respect to employment in particular, politicians can use public employment as a way of supporting their constituents including through handing out jobs as well as using other preferential treatment, such as contracts. Consequently, the public sector – of which the SOEs are a part, of course – has been an important mechanism of preferment, particularly when the taste for socialism was in the ascendant. Even in the present day, SOEs in China are explicitly viewed by government and their oversight agencies as loci for stabilizing – and sometimes creating – jobs. In the immediate aftermath of the COVID-19 crisis in Wuhan, the Chinese commission supervising state-owned assets (SASAC) announced that it was the role of SOEs to create jobs and ensure no job losses.[35] Take the example of Weliangye Group, a major liquor company that is a SOE whose headquarters are in the province of Sichuan. The company's CEO promised that no layoffs would happen nor would there be any reduction in wages for its workers. In addition, the company announced a series of investment projects aimed at boosting employment and supporting up to 300,000 jobs in its up- and downstream production and distribution chains. Other SOEs – such as Poly Group, a company with very diverse interests ranging from military defence sales to an art auction house – similarly announced additional projects aimed at creating new jobs. At the same time, SASAC declared that private companies should also ensure that employment remained stable. In sum, the employment level is both a highly political and a highly sensitive matter and no more so than in those parts of the economy where government can exert a direct influence on decision-making, whether in companies or the public administration.

There is also a wider process at work, albeit one that is particularly accentuated with public sector entities. The strategic place of business groups and large companies – both private and publicly

---

[35] www.sasac.gov.cn/n2588020/n2877938/n2879597/n2879599/c14935790/content .html.

owned – is closely related to their scale and it is the larger firms and groups in terms of both revenues and employment that tend to be the most politically connected businesses. Research focused on Asia and elsewhere has shown that firms connected to politicians may expand in size but not in productivity.[36] What this also implies is that many connected companies – particularly state-owned ones – may actually be larger in terms of employment than they need be from an efficiency perspective. Part of the reason that they grow larger than they should is, of course, to preserve, if not enhance, their ability to exploit political connections. Their owners – in the case of private firms – may well be reluctant to trim jobs lest they antagonise their political and other connections. Even so, the incremental employment that has resulted pales in comparison to the loss of potential jobs that a more competitive business environment and market structure would permit. In addition, excess hiring or retention of workers ensures that political connections and preferential treatment induce downward pressure on productivity.[37] Placed in broader context, such consequences only compound any adverse impact on productivity that result from other reasons such as any misallocation of capital or labour.[38]

6.7 PULLING THE PIECES TOGETHER

A central aim of all Asian governments – as well as the many development agencies and associated advisers – has been to increase the amount of formality and by extension, the share of more productive and remunerative jobs. All of this has sometimes been rolled up into the slogan of creating 'good' jobs.[39] But whatever that exactly means,

---

[36] Bussolo et al. (2019); Akcigit et al. (2018).

[37] Huneeus and Kim (2020) argue that increased size due to lobbying in the United States may result in aggregate productivity being lower by as much as 11 per cent.

[38] Several decades ago, productivity in China and India was estimated to be lower by around 50 per cent compared to the United States due to misallocation of factors. Subsequent policies will have reduced levels of misallocation but not eliminated them. See Hsieh and Klenow (2009).

[39] As for example, in the World Bank's 2013 World Development Report.

it is evident, sadly, that this has mostly failed to materialise. The reasons why many SOEs have performed so badly – including weak governance and politicised decision-making – are well documented. What is more perplexing is why has the organised private sector not been able to lead a stronger employment charge and create a cascade of good, formal jobs? Although it has been argued that this is because the public sector has been too large and diverted needed resources, the argument has become less convincing over time as the number of SOEs has tended to decline and expansion in the size of the total public sector has stopped or even been reversed. Rather, a more convincing approach is to turn to the ways in which private companies and business groups conduct their business, the incentives they face and the wider consequences of their actions.

In Chapters 3 and 4, we highlighted the ways in which Asian business groups and others leverage political and other connections to their benefit. These actions are mostly aimed at securing advantage and, where required, entrenchment. They sometimes involve the need for reciprocity and that may include creating jobs for their connections or jobs for the constituents of those connections, or, equally, ensuring that workers do not get laid off even when lay-offs would be warranted. Indeed, a common feature of the connections world is that connected companies and business groups are actually rather good at protecting jobs. The latter are particularly dexterous in insuring their constituent companies and affiliates against risk. This is done by transferring resources internally so as to prevent bankruptcy and/or job losses in poorly performing parts of their empires. Whilst this may not be efficient, insofar as it transfers resources from high to low productivity users, it does ensure that business groups are able to respond to the demands of politicians and others – including their employees – who favour stability in employment. For instance, as COVID-19 began to grip the Indian economy, the country's emblematic business group – Tata Sons – did not make any workers redundant, despite shuttered factories and collapsed demand. Many other business groups across most Asian economies have gone out of their way to contribute to government attempts at coping with the pandemic, including by maintaining employment.[40]

In addition, while many business groups do compete hard with each other, many activities, and sometimes sectors, are dominated by incumbents. The result, as we saw, is substantial concentration and market power. Such consequences carry good news for those fortunate enough to secure employment in these companies as they, in effect, form a relatively privileged layer of the labour market. Their jobs are, by local standards, filled by relatively well-educated and skilled workers who in turn are well paid and who tend to have some degree of job security.[41] Their management also benefits. In Malaysia, for example, business groups reward their top management (often including family members) far more generously than their comparators or reference to the equity market or performance would appear to warrant.[42] In South Korea, family business groups have tended to reward family members at levels superior to non-family professionals.[43]

Whilst in the past, the ladder to a good job in Asia was often considered to be through the public sector, the ladder that exists today is mostly positioned besides the organised private sector and those business groups that form such an important component. However, that ladder is narrow and the those who can climb up it are far from numerous. Although those who do may form part of a modern 'labour aristocracy', it does not appear that insiders have the ability to limit hiring in order to preserve their privileges. Although trade unions are present, much of their power has historically been concentrated in the public sector as, for example, in India. In sum, the connections world

---

[40] For instance, many of the main Filipino business groups (inter alia, the LT Group and San Miguel as well as the Sy family) have donated resources – financial and material – to the health system.

[41] This extends to the management of many business groups whose compensation tends to be attractive, particularly when that management is a member of a controlling family. In Malaysia, for example, in recent years BG CEOs have received double what would appear to be implied by either profitability or market capitalization. In contrast, CEO pay for government-owned or -linked companies has been far more restrained. See *The Edge*, Singapore, 24 June 2019.

[42] Special Report: *The Edge*, Singapore, 24 June 2019.   [43] Kim and Han (2018).

is part of a wider nexus that holds back the proliferation of good jobs, restrains broader productivity growth and sustains pervasive informal economies. But the consequences of the connections world run wider, not least when considering the potential for these economies to develop modern institutions for managing labour market risk, notably the threat of unemployment.

6.8 SOME POLICY CONSEQUENCES

6.8.1 *Connections versus Social Insurance*

We have seen how the Asian economies are marked by relatively small, but privileged, formal sectors. Aside from entrenching some companies and business groups, privileges also accrue to those that are employed in these companies. Those benefits stretch from relatively high wages to job security. In addition, their employees commonly have access to benefits provided by either the state or the company. Yet, even if significant numbers of citizens have access to benefits or transfers that, in one form or another, aim to address poverty, in most of Asia very few workers have access to benefits that provide income to a person who loses his or her employment.[44]

In the rich world, support for those involuntarily losing their jobs mostly takes the form of unemployment insurance where loss of work triggers the payment of a benefit, either as a share of the previous wage or as a flat rate. When such systems exist, they are normally funded by contributions from a combination of employers, employees and government. Of course, such systems are mostly infeasible in poorer countries, both for reasons of affordability but also the lack of institutional capacity to implement. As a consequence, most developing economies, including in Asia, tend to provide severance payments to workers in the formal economy when they lose their jobs and, for the most part, nothing to other types of workers. Among the wider implications of the fact that employment risk sits squarely on the

---

[44] As documented by the World Bank (2020a) in its Aspire database.

shoulders of most workers are the high savings rates by households as they attempt to build up reserves for such eventualities (including health risks). This is widely acknowledged to be the case in China, but it is actually present throughout Asia.

Despite unemployment benefits being only available to a very limited number of workers in the formal economy, citizens in Asia actually have growing expectations about how the state should mitigate the risks that they face. For example, responses to the World Values Survey show that Asians increasingly view protection against job loss and other employment risk to be an important and a desirable feature of public policy.[45] Even so, social insurance and social security systems are largely missing or have trivial coverage. The exceptions are the richer Asian economies, such as Singapore and South Korea, which have some of the elements of social insurance that are found in Europe or Japan.

The limits of existing social insurance are well illustrated by the situation in China. There, social insurance is available for a limited group of workers employed in SOEs and some private formal sector companies. This has meant that when job losses have hit, few have any support to fall back on, other than their savings and family. For example, at the height of the COVID-19 pandemic in March 2020, independent observers estimated that as many as 130 million people, or 30 per cent of the Chinese urban labour force, was out of work or on furlough (even if official statistics reported an unemployment rate no higher than 6 per cent). Whatever the real rate of unemployment, in the first three months of 2020, only 2.3 million people received unemployment benefits and these only averaged $190 per month. To put this in context, the average monthly urban wage at that time was around $1300.[46]

China's fairly low levels of unemployment insurance along with the limited coverage are closely related, of course, to the fact that

[45] As, for example, suggested by responses to recent rounds of the World Values Survey.
[46] The Economist (2020a).

employers exercise a great deal of caution in laying off workers. With SOEs in particular, the authorities and management have consistently shown a low tolerance for firing workers on any scale. Instead, voluntary separations and early retirements have often been used. Although, the reluctance to fire workers does not extend to the same extent to the formal private sector, larger companies – particularly those with close connections to government – have to proceed very cautiously with any redundancies at scale. What this all implies is that the main benefit for workers in the formal economy remains strong de facto job security.

One consequence of these arrangements is that, in effect, job protection and unemployment benefits are more like substitutes than complements. Put differently, when job protection is the main way of minimizing unemployment, governments feel it less pressing to create broad-based and broadly funded systems of unemployment insurance. And although there may be some merits to retaining workers rather than firing them, the absence of robust options for those who lose their jobs means, of course, that there will be less restructuring and more persistent inefficiencies than would otherwise be the case. Once again, the process of creative destruction is suppressed. As a parenthesis, the preference for job rather than worker protection is by no means unique to China. The accompaniment is an inability to move towards an institutional system that can effectively deal with failures. This inflexibility testifies to a mismatch between the dictates of a market economy and the imperatives of an autocratic system that prefers to use discrete and often rather personalised levers for achieving its aims, as well as its high degree of sensitivity to changes perceived as destabilizing.

Despite the fact that very limited benefits are available to most workers, ironically, China's system does impose high taxation on labour income for workers in the formal parts of the economy.[47]

---

[47] This taxation decreases with income so that the impact falls more heavily on lower wage workers.

What economists call the tax wedge – the ratio between taxes paid by an average worker and the total cost of labour for an employer on average – is fairly high at around 32 per cent. To put this in context, most European countries have a wedge ranging between 40 per cent and 50 per cent, while South Korea's is 23 per cent and Indonesia's is below 8 per cent. Even if, on balance, high rates of labour taxation – as in China – have not encouraged companies to lay-off workers, these rates serve as a further disincentive for private companies to create jobs in the formal economy.

Although the absence of social security systems is far from unusual in most emerging economies, given the income levels of some of the Asian countries, along with the changing expectations of their populations, it raises questions as to why some enhanced form of coverage has not naturally emerged. Not only has this remained largely off the political menu but there has also been surprisingly little discussion about other forms of income support, such as universal basic income.

So, what accounts for this unwillingness to put in place modern systems for abating risk? Part of this is undoubtedly related to the cost of such insurance and to the belief that most Asian economies are insufficiently wealthy to be able to fund these sorts of programmes at meaningful levels of income support. Yet, in the European Union the cost of unemployment related benefits (which is broader than just unemployment insurance) has in recent years been below 1.5 per cent of GDP. For example, France spends a bit under 2 per cent, Germany around 1.4 per cent and the United Kingdom 0.3 per cent. For Asia's middle-income economies, it is quite conceivable that they could sustainably allocate magnitudes of spending comparable to, or not far below, these shares of GDP.

A second argument is that such programmes are actually undesirable, insofar as they can promote welfare dependency and create adverse incentives for workers to search for jobs. This reasoning has certainly been widely aired – and sometimes actively espoused – in China. Yet, the evidence from many countries' experience is that such disincentives – which clearly can be created – can also be mitigated through good design.

But there is a further reason of equivalent, if not greater, importance. That is, in the connections world, neither government nor companies actually have a strong interest in promoting arms-length methods of dealing with labour market risk. They prefer, rather, to rely on the discretion and associated bargaining over employment that we have already discussed. Jobs can be created, their destruction tempered or halted as a result of interactions or even haggling between politicians and employers. Further, the bulk of workers, namely those who function in the informal economy, lack neither the bargaining power nor connections that might help sway politicians. This implies that the bargaining over employment is largely, if not exclusively, over the privileged parts of the economy. This balancing act comes, however, at the expense of those who live and work outside such privileged parts.

In sum, to deal with the periodic shocks to companies that market economies throw up, the Asian economies have preferred to use firms – particularly SOEs but also larger formal sector private companies – sometimes by governments exerting suasion – as the principal places for dealing with employment risk. The architecture of containment has always tended to favour postponement of difficult adaptation and job losses. As such, politicians have been far from shy in using companies to act as forms of shock absorbers and companies in turn have accepted this as part of the necessary cost of being connected. In the case of business groups, they often internalise these complex objectives through the use of transfers or forms of cross-subsidy to weaker parts of their empires. But it is also the case that to cover the consequences of inaction, public subsidies – explicit or implicit – have commonly provided some of the required financing.

6.9 CONCLUSION

Asia has surely made huge strides in creating employment for its citizens and in improving the income prospects of its citizenry. An earlier predilection was for the public sector to act as the main motor of job creation. Yet, over time this motor has stalled. Governments have increasingly relied on the private sector to take up the baton. But, even with declining birth rates, the pressure to increase the number of jobs will remain pressing over the coming decades. Those pressures have been accentuated by the onset of early de-industrialisation,[48] which has begun to restrict the scope for labour-intensive manufacturing's ability to create jobs. Further, the fact remains that the great majority of jobs in Asia are still in the informal sector and are precarious. The connections world and other larger companies, including foreign-owned ones, have certainly created new and productive jobs. But they still account for rather low shares of formal, let alone, total employment. Moreover, the core levers of dynamism in a market economy that involve the entry and exit of companies have been held in check. Business groups and other established companies may compete with each other but the scale of entry and rivalry is limited. Moreover, most companies fail to navigate an upward path, whether denominated in terms of jobs or revenues. This results in a pyramidal structure in which a limited number of medium-sized firms, and above them a sliver of large firms, sit on a large base of small, often informal and mostly low productivity firms.

With regard to public policy concerning employment, what we find is that there is a continuing tension between using levers of influence or control to create employment for, and in, connected entities as against providing a supportive framework for the wider enterprise sector to create productive jobs. In general, governments have preferred to stick with wielding discretionary policies rather than promoting broader job creation. This has resulted in the structures of employment that have been described in this chapter, structures that have proven to be very long-lasting. As such, the pressures for the creation of more productive jobs still remain omnipresent. And these are only likely to be accentuated by the sorts of accelerated

---

[48] Rodrik (2015).

technological change that have emerged in the advanced economies and are increasingly spilling over into emerging Asia. Among other consequences will be a higher amplitude of underlying fluctuations in the demand for labour and hence for employment. This is likely to be a problem for governments that see such fluctuations as challenges to their core constituencies (including the connected companies) but also to their political acceptability by the far more numerous workers in the informal – and yet more precarious – economy. We return to discussing these tensions in the next and concluding chapter.

7 Whither Asia?

Prospects and Policy Challenges

7.1 A SUMMARY OF OUR ARGUMENT

We start this chapter by summarising the main arguments and conclusions of the book to this point. After that, we turn to discussing some of the policy changes that might help Asia make further progress. For sure, Asia has made huge strides. Although some countries have grown and developed more than others, there has been a generalised improvement in living standards and prospects. That improvement is palpable and impressive. Placed in perspective, barely half a century ago, Asia was still viewed as the repository of vast amounts of poverty and insufficiency. Now, some herald the dawn of a new Asian century of prosperity. Whilst the hyperbole may have been inverted, it is quite evident that something radical has occurred.

Our book has provided a fresh perspective on Asia's success. It has concentrated on the central characteristic that is common across the countries. That is, the pervasive role of networks and connections, both within and between business organisations and between the corporate sector and the body politic. We have argued that while the connections world has played a role in the economic success of Asia over previous decades, it also has a number of undesirable features that have increasingly come to the fore and which bring into question the sustainability of that success.

Part of the reason for why the future path may be more rocky lies with the nature of the growth model that has been pursued over much of the past half-century. In that time, extensive growth – the deployment of large amounts of capital and labour – has been the main driver at work. Although productivity has undoubtedly improved – especially in East Asia – those improvements have generally been less significant, as well as being quite variable over time and place. Even if the path of extensive growth is not wholly exhausted, it has less far to run as marginal returns to factors shrink. At the same time, in some places – notably in much of China – an easy demographic dividend, realised through large-scale migration of workers out of the rural economies into urban areas, has pretty much run its course.

Yet, it is not just the need to raise productivity and go beyond extensive growth that is the challenge. Chapter 3 described in detail how many of the foundations have been laid on the basis of connections and the networks associated with those connections. These run between the body politic and business as within and between businesses. Although we showed that there are some significant differences between countries in how such connections are configured and their composition, these networks of connections are ubiquitous and deep and provide a common framework for interaction. Further, they are nothing if not resilient. As Asia has developed, the connections world has adapted while continuing to entrench. The transactional content of connections has continued to dominate along with the need for reciprocity between players in the connections world. For example, established businesses – particularly the family-owned business groups that are such a strategic component of the Asian world – are able to conserve their market power by getting politicians and officials to bring the rules and regulations of the state to bear in ways that preserve their market dominance. Large business groups and their component parts have thus been able to entrench themselves even more deeply behind – among other methods – the protection of domestic industry as well as barriers to entry including through subsidies and favouritism in procurement, contracts and the provision of soft loans by state-owned banks.

This mutually supporting connections world has as one of its lynchpins the family-owned business group. Such business organisations are well adapted to the reciprocity game. In Chapter 4, we saw how their opaque ownership arrangements and their non-transparent accounting processes facilitate legal, semi-legal and outright dubious relationships, both with other firms and also with politicians and civil servants. Moreover, their structures ensure that the development process acts disproportionately to enrich the oligarchic dynasties that own them. Most importantly for the future of Asia, these business groups are often highly entrenched across the bulk of resource, capital and product markets in the region. While their scale and diversity may have made them important mechanisms to mobilise resources in the early phase of economic development, they have now become the vehicle whereby the connections world can impede the next stage of development.

This is clear when considering innovation – a critical ingredient for future growth. Chapter 5 evaluated how well placed the Asian economies are to innovate. We found that in most of Asia, new technologies and products are sourced from abroad, and domestic firms are primarily engaged with adaptation and diffusion. The ability to mobilise resources through concentrated business organisations has facilitated this process, even if the connections world means that such innovations mostly favour incumbents over new entrants. But a key question is whether the Asian economies can move to undertaking innovation on their own account. This has certainly been the aim of the education and science policies that have, in particular, been adopted in three countries – China, India and South Korea – as part of the drive to create national innovation systems. While the rhetoric has often hailed victory, especially for China, the actual outcome remains much less settled and there are sound reasons to err on the side of caution. As ever, the main issue concerns whether Asian business groups can go beyond rent-seeking to spearhead innovation. While support from the state can certainly spur innovation, including in business groups, the incentives to innovate will be compromised when the businesses meant to innovate have dominant market positions and where their market power allows them to erect barriers to entry by innovative new ventures. These barriers are of course made even harder to overcome when supported by regulations and subsidies through political connections. The political system – as in China – also affects the ability to innovate. For instance, the freedom to innovate and experiment in the business sphere – indeed the scope for human creativity more generally – may also sow expectations of parallel freedoms in the political sphere, which the political authorities dare not tolerate. The Chinese artist, Ai Weiwei, puts it pithily, 'when human beings are scared and feel everything is exposed to the government, we will censor ourselves from free thinking'. And, recently, it is also becoming clear that even the type and contour of business innovation has to be subject to state direction. That appetite for control has not historically been propitious for sustained innovation, as the experience of the Soviet Union testifies.

A further serious, and even more immediate, challenge posed by the connections world in Asia is to create enough jobs to employ their large labour forces. Chapter 6 showed that while business groups are very successful at generating profits, they have been far less successful in creating jobs. Dual labour markets have emerged with segmentation, whereby good earnings and prospects for those employed in the formal sector – mostly business group affiliates or SOEs – coexist with a far larger mass of poorly paid and precarious jobs mostly of an informal nature. Yet, in the next twenty years, Asia will have to create a further million jobs a month, even without considering the likely impact of technical change, including the accelerated use of robots and artificial intelligence. The creation of very limited numbers of productive jobs is intimately linked to the market dominance of business groups. Entry barriers and other restrictive regulations limit the incentive for smaller firms to become formal and to grow. As a result, Asia differs from North America and Europe through the absence of middle-sized firms creating jobs and growing into organisations able to compete with the larger incumbents.

With this background, the obvious questions that need asking concern the ways in which Asia can move forward and address the strong entrenchment of the connections world and its main players. As such, this chapter is not about prognosticating on the likely path of each economy covered in the book. Rather, it first evaluates the scale of the challenge as indicated by a measure of the intensity of connections taking into account the nature of the political system. Autocratic regimes tend to have a greater intensity of connections which suggests that they will be less likely to evolve naturally to a less concentrated structure. It then moves on to propose a series of policy shifts that could roll back market dominance and entrenchment. Needless to say, coming to grips with business groups and their symbiotic relationships with politicians is a massive task, not least because the current equilibrium appears to be successful and benefits both politicians and business. Even though there may be negative effects on future growth, no one has an incentive to rock the boat.

Since the barriers to change are deep and numerous, we propose a multi-pronged approach to transforming the connections world. First, we suggest a series of measures to inspire the gradual disappearance or transformation of the business group into more transparent and better-governed business organisations. In particular, we suggest policies covering corporate governance, pyramidal ownership structures, mergers and cross-holdings, inheritance taxes and new types of competition policies. We also propose limits on the discretionary scope of politicians to leverage their connections combined with stronger enforcement.

We conclude the chapter by revisiting the remaining fallibilities of the connections world introduced in Chapter 1. We see both the jobs challenge and inequality as major pressure points for the future. We end by putting Asia in a wider historical context while also reflecting on Asia's role in the paramount policy matter of our times – climate change.

7.2 THE INTENSITY OF CONNECTIONS AND IMPLICATIONS FOR THE FUTURE

If the connections world is so important, a sceptic might argue that the ubiquity of networks and the lure of connections transcends Asia. Just consider the response to the COVID-19 pandemic. For example, the United Kingdom chose to give priority to close connections of politicians in the rapid allocation of contracts for protective clothing and other healthcare purchases. The rationale for bypassing normal procurement channels was simply urgency. The outcome was the granting of contracts to companies or individuals on the basis of whom they knew, with doubts also about the terms of those contracts.[1] Yet, if this example illustrates the strength of the underlying pattern of behaviour, it is, however, of quite a different order of magnitude both in scale and scope to the vibrant connections world characterising Asia. In Asia, the cornerstones of the connections world that account for its resilience are the pervasive concentrations of economic power. Particularly in the Asian autocracies, concentrations of political power have also facilitated – and sometimes directly fashioned – the translation of connections into economic standing. This is blindingly obvious in the case of the dominant Chinese tech companies. But it would be thoroughly misleading to think that autocracy is a precondition for embedding the connections world. Very similar ways of operating – and consequences – have characterised the different variants of democracy existing in Asia. The main difference has been that in democratic settings the risk of political turnover and loss of advantage has both acted as some sort of constraint, but also induced a variety of hedging strategies to temper such risk, including maintaining highly diversified family-based business groups.

7.2.1   *Variations in the Intensity of Connections*

Even if Asian economies have a common salience of the connections world, there are, nevertheless, some significant and strategic differences across countries that matter when considering future prospects. A difference that matters concerns what we can term, the intensity of connections. As we saw in Chapter 3, connections – and the numbers

---

[1] www.nytimes.com/interactive/2020/12/17/world/europe/britain-covid-contracts
.html?referringSource=articleShare.

of nodes and edges that summarise the underlying networks – vary substantially in number and coverage across Asia. This is hardly surprising, not least because of the very differing sizes of these economies. Thus, for example, massive countries – India and China – have a relatively low number of nodes and edges per capita when compared with some of the smaller countries. But that does not mean that their connections worlds are less potent or influential. Indeed, as the earlier chapters have shown, this would not be a sensible conclusion to draw.

There are some other components to intensity. The first concerns not just how common connections are, but how close are the relationships between businesses and politicians. In other words, there may be pervasive links between businesses and politicians, but these may be spread over large numbers of actors. By contrast, in some jurisdictions, those connections – and the ones that really matter – are crystallised around a limited number of players. These players are likely to occupy strategic positions in the networks. In most Asian contexts, such strategic locations are mostly occupied by businesses that are themselves networked through their format as business groups.

What is also suggested is that the intensity of connections will reflect the weight of particular sectors and parts of the economy where connections play a central role. These are almost entirely in the formal economy. The large informal and unregulated parts of these economies – despite accounting for much of employment – are in effect excluded from the connections world. In turn, the intensity of connections and their locus in the formal economy is closely related to the extent of economic concentration and, with it, the degree of market power that connected entities possess.

7.2.2 Economic Concentration and Intensity of Connections

An accessible way of putting these elements together is simply to relate the intensity of connections to the extent of economic concentration in a country. This is done in Figure 7.1 which summarises our view of these different Asian economies. Not only is it obvious that

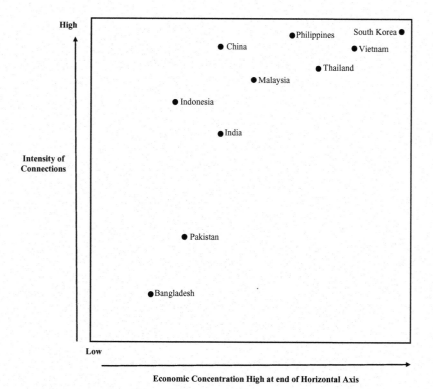

FIGURE 7.1 Intensity of connections and economic concentration

there is a good deal of variation but there is also a clear underlying association between political system and intensity. In particular, autocracies tend to have very high intensities of connections – hardly surprising given the central role of the unique political party and the extent to which political imperatives influence the direction of economic policy. However, high intensity is not a monopoly of autocracy.

7.2.3   China and India

Moving to specifics, what of the two giants: China and India? The number of connections and the habits of linking politicians and businesses are certainly well embedded in both places, albeit in somewhat different loci. For example, the part of China's economy that is explicitly state run is far larger than in India. And, as we have seen, the tentacles of public policy and preferences have embraced significant parts of the private sector, not least the leading Chinese tech firms. As a consequence, China has a very high intensity of connections, while economic concentration is high in particular strategic sectors. India has a far lower intensity of connections, but its extent of concentration is broadly similar to that of China.

Yet, in both instances, what their size camouflages is that parts of the economy – in the Chinese case, whole sectors, in fact – are marked by a very high intensity of connections and an associated dominance of a small number of powerfully connected players. For instance, in earlier chapters, we have talked about how three to four business groups dominate the Chinese new tech economy. Each has unambiguously strong ties to government in numerous respects. In other parts of the economy that currently attract less attention and figure less prominently in strategic policy making – such as steel making – a limited number of state-owned enterprises are also dominant. These forms of legacy investments still carry influence, not just because of the role they play in infrastructure development but also because they employ substantial numbers of people.

In India too, there are some sectors where specific business groups have continued to carve out major stakes and translate their connections for their advantage. Connections are still central in activities where government permits or concessions are involved. However, as Chapter 3 made clear, India's attempts at liberalisation since the 1990s have not just opened up space for more competition but have also affected the calculus for business groups to stay diversified. Indeed, whilst the power of connections has far from evaporated, it has become less salient than it was thirty years ago.

7.2.4   Philippines and Thailand

Away from the giants, there are clear instances where intensity and concentration are tightly correlated. The Philippines is one such case.

The country has a very high intensity of connections. Not only are there strong and pervasive links between businesses and political players but the dominant businesses are themselves compressed into a remarkably small number of business groups and interests. These groups are in turn controlled by a very small number of extremely wealthy families. Needless to say, this translates into strong and persistent concentration in markets, let alone in influence and power.

High intensity of connections and its translation into high economic concentration is also very much present in Thailand. While many of those connections have remained undisturbed and some incumbents have succeeded in entrenching themselves even further, there have been conflicts within and between the connected. Such lateral pressures (as opposed, say, to distributional tensions from 'below') have manifested in struggles for market sway or other forms of preference. In the Thai case, disputation that was originally mainly about how the contours of connections were set and applied became transformed into a broader political dispute that, ultimately, shifted into conflict on the streets, let alone in the courts and parliament. These have been sufficiently severe that the country has been tipped into political turmoil, including into autocracy. What this illustrates are the ways in which rivalries among the connected – particularly among the large business groups – may not necessarily be containable, spilling over into wider channels that can be highly conflictual. In other words, a high intensity of connections and associated economic clout does not necessarily amount to a stable equilibrium.

7.2.5   South Korea

For a stable system based on high intensity and concentration, we should turn to South Korea. Without exaggeration, much of the connections world is funnelled through, and into, a very small number of massive family-based business groups, who in turn dominate much of the economy. But – as Chapter 5 indicated – there has been a growing sense over the past decade that such concentration also carries major risks, not least by failing to allow a flourishing of smaller, more entrepreneurial, innovative companies. The state has attempted to refashion its approach, including by trying to create a more decentralised and smaller-scale world of new firm entry and growth. Other measures have been taken to rein in the larger business groups, but the results have as yet been held back by the power of incumbency and the continuing intensity of connections.

Even so, South Korea has several advantages in trying to steer this path forward. The first is that – despite endless cases of corruption and influence peddling involving politicians and the main business groups – its political system is that of a pluralistic and competitive democracy. Turnover of parties and an independent judiciary have increasingly placed limits on the abuse of power and privilege. The second advantage is that although the incumbents have benefitted from significant financing from the state, most of the sectors and activities with which they are involved are open to external competition. Consequently, although South Korea exhibits the features of a highly concentrated form of domestic capitalism, it has been one that has, over time, come to rely less and less on protection from external competition. The discipline that this creates mitigates against underperformance and, to some extent, a taste for providing public subsidies.

7.2.6 Reducing the Scope for Connections in Different Political Systems

Shifting away from high intensity of connections is likely to prove to be far more problematic in China and in Vietnam. That is, of course, to do with the fact that those connections are absolutely integral to the way the wider political system is constructed. Although the contours of these connections have changed through the emergence of very large and often very profitable privately-held companies, these – almost invariably – have very close ties to government and Party. In those sectors deemed strategic, the dominant private companies receive substantial financial and other support from government – not least through protection from external competition.

The intensity of the connections is, as a result, very high. For those businesses that benefit, the Faustian pact that they have done with government and Party ensures that support equates with pliancy in areas where political prerogatives are pre-eminent. These companies may be private but there are clear boundaries that they should not cross. The price of support is concordance with the government's broader objectives.

The Chinese or Vietnamese states have a wide range of possible sanctions that they can apply against those who are deemed to be out of step. In 2020–21, the tempo of regulatory assault against major businesses in China has been raised as the state has tried to enforce greater control. However, some options – such as forcing out management or changing governance structures – have the demerit of disturbing carefully crafted systems and their underlying commitments. The paths for adaptation, involving a loosening of the intensity of the connections world, are both less wide and less susceptible to trial and experimentation than in a democracy. The very strengths that highly intense connections and concentration can bring – such as economies of scale, resource availability and protection from competition – are, of course, also capable of being sources of fragility and disruption. The absence of countervailing forces and a broader ecosystem of companies that more decentralisation would have conferred poses a challenge to future innovation and competitiveness, points that we emphasised in Chapter 5.

The challenges of the connections world facing India are rather different from China as a result of the characteristics of the political system and the directions of change. Since the 1990s, policy has aimed to facilitate greater competition and to break down the iron clasp on particular sectors or activities that incumbent business groups had accumulated in the decades following the country's Independence in 1947. In Chapter 3, we highlighted how a combination of liberalisation and the emergence of new activities in areas of the economy largely untrammelled by the dead hand of regulation has led to some dilution of the power of incumbency. Some incumbents have successfully adapted to greater competition, others have seen their relative standing wane. Encouragingly, these combined developments have shaken some of the roots of the connections world. But, in a sign of its continuing resilience, a number of business groups have recently and successfully parlayed their connections to the current crop of federal and provincial governments to their advantage. Probably the starkest example is one that we mentioned earlier, namely the Adani Group.[2] Even if less pervasive than previously, this emphasises the need to put in place more effective policies aimed at improving both oversight and transparency, but also to addressing the principal consequence of the connections world: the attenuation of competition.

7.3   DEALING WITH THE DARK SIDE OF BUSINESS GROUPS

Most Asian economies present a stark dichotomy between the almost cut-throat levels of competition that exist in parts of the economy – particularly in the informal economy – along with the less competitive, sometimes collusive, parts occupied by the big business groups. It is not that these groups do not compete with each other. They do. And often fiercely. But it is that the markets in which they compete are mostly marked by a very limited number of players. So, while monopoly is rare, other variants of market power are not. Further, as we saw in Chapter 4, being a part of a business group can confer overall concentration, not least because business groups allow resources to be transferred internally in ways that permit them to assemble huge firepower in any particular market. Needless to say, those resources also allow them to line up required support from the political networks to which they are attached. What can be done to induce greater competition and to provoke more entry and exit of companies?

---

[2] Connections appear to have delivered airport concessions and regulatory indulgence while facilitating continuing access to bank and other finance, despite very high debt loads. See *Financial Times* (2020).

The answer necessarily lies in dealing with the business groups that sit at the centre of the connections world web. Most of these business groups are dominated by one or a few families. Their origins often date back to when there was a strong need for company finance, capital market institutions were weak or non-existent and trust in external institutions was frail. Successful families derived cash flows from solid businesses – in trade, resources and manufacturing – and then moved on to absorb other promising companies – sometimes the fruits of their connections to politicians – across a diverse set of activities. In due course, they created large, sprawling, interconnected businesses and financing arrangements. These have generally persisted to this day. Even newer entrants have tended to mimic the same broad format. That format has permitted extensive use of internal funding transfers within the group. The channels include, dividend payments, intra-group corporate loans as well as tunnelling resources between businesses or using transfer pricing to charge for goods and services between group members at prices which transfer profits from firms that retain surplus resources to companies that require additional funds. At the same time, to address possible shortfalls in funding, many business groups have taken on pyramidal ownership structures. This has allowed raising additional resources from minority shareholders whilst ensuring that the family retains effective control through cross-holdings and other mechanisms.

While such structures may address corporate financing needs, they come at a high price including lack of transparency and very opaque corporate governance. At the same time, the economic and financial consequences of the business group format are both complex and, not infrequently, adverse, at least at the level of the economy. Business groups often stand in the way of creative destruction as they tie up resources by allowing inefficient firms to survive. Moreover, their size and diversity has allowed Asian business groups to accumulate significant market power which, in turn, has been used to restrict the entry or growth of new entrants and, ultimately, limit innovation. Barriers to entry also help preserve market dominance and profitability which can then be used to entrench further. This business format also explains why, despite the fact that most formal employment in business groups is secure and relatively well paid, the bulk of employment still falls in the informal sector who remain locked out of more lucrative – and more productive – markets.

This is not the sum of it, as powerful, family-based business groups commonly survive and grow because of their close and symbiotic connections to politicians and government. Business groups can provide politicians with a host of supportive resources, including financial, but also through media support due to their ownership of TV, newspapers and online companies. They can provide jobs and favours to their political supporters and voters and are not averse to funding major public initiatives if they see payback. The politicians' side of the bargain can also deliver significant benefits, such as restricting or preventing competition – including foreign competition. They can also favour business groups in state procurement and erect a barrage of regulations that choke off entry and buttress incumbents. They can direct, or turn a blind eye, to state-owned banks funding business groups, including on terms that are highly preferential and they can use state-owned utilities to subsidise favoured companies in the pricing of, inter alia, electricity or transport. They can channel the fruits of state-led invention towards innovation by business groups. As our book has shown, the menu is not just long but often highly rewarding for its beneficiaries.

What all of this implies, of course, is that the incentives to move away from business groups and to modify behaviour are tempered, if not entirely offset, by the mutual benefits of the current systems in place. The costs of being a diversified business group which might have been expected to increase as the Asian economies modernised and capital markets deepened have proven far less cumbrous than many have expected.

Consider the matter of the cost of capital. In advanced economies, diversified firms, even with relatively transparent ownership arrangements and governance structures, suffer from a significant discount in their valuation, in the region of 10 per cent.[3] This makes it more expensive for firms such as business groups to raise funds on equity markets. In the emerging market world, the evidence also points to a large and systematic discount for diversified firms of the order of 7–9 per cent.[4] In principle, this discount should provide the controlling families with a clear and strong economic incentive to break up their business groups.

But realities are different. Business groups in Asia are for the most part expanding in size and scope, and their share of economic activity is tending in most places to increase rather than fall. Among the reasons are their ability to continue to leverage the connections world, including in securing funding on favourable terms from state-owned lenders and other institutions susceptible to political pressure, rather than by recourse to equity markets.[5] Furthermore, business groups have access to substantial financial resources within their own operations. In India, shifting resources around or tunnelling occurs not through the transfer of operating profits, but through non-operating profits, such as writing off debts.[6] Indeed, a major way Indian business groups allocate resources is through intra-group loans which can always be subsequently written off. Group loans are also used to finance profitable investments.[7] Business groups are also not averse to generating funds by expropriating minority shareholders and by using related party transactions in which transfer prices are used to shift profits from one firm to another.[8]

Putting all this together suggests that the continued survival and growth of business groups is not mostly due to higher productivity, but their ability to provide their affiliates with important 'non-market' advantages, such as access to preferential finance and export markets through their political connections, protection from the

[3] For example, Lins and Servaes (1999).
[4] Lins and Servaes (2002); Khorona et al. (2011).   [5] Crabtree (2018).
[6] Bertrand et al. (2002).   [7] Gopalan et al. (2007).
[8] Baek et al. (2006); Goldman and Viswanath (2017).

predations of discretionary regulations and preferential routes to resolving issues arising from competition or merger policy.[9]

These attributes provide an answer to the obvious question of why these particular patterns of business organisation and behaviour have proven so persistent in Asia. In short, the logic of market entrenchment and political connections in practice outweighs any marginal disadvantages of the business group structure. The connections world is often hugely beneficial to the principal players, whether businesses or politicians. Neither side has incentives to deviate from, or repudiate, these relationships and the networks on which they are based, even though there may be adverse long-term consequences.[10]

7.4 BREAKING THE TIES THAT BIND

In a context where players in the connections world lack incentives to change how they behave, what if anything can be done to shift these incentives and help force change? As we shall see, this task is by no means straightforward: the barriers are numerous and the power of incumbents cannot be underestimated. Even so, there are two angles for possible intervention that are particularly relevant. The first concerns the business group side of the connections world and, specifically, the incentives for the continued prevalence of family-based business groups, notably those with pyramid-like structures.

The second concerns applying pressures on politicians and political parties to limit the discretionary scope for leveraging connections with business and businesspeople. This involves increasing not just transparency across a range of possible transactional loci, but also creating a workable system of enforcement. Needless to say, these are aims that fit with democratic societies far more than in autocracies where discretion and opacity is the norm.

Let us turn to the first. Looking at historical experiences of reining in or dissolving business groups suggests that it can be feasible

---

[9] Tihanyi et al. (2019).
[10] Economists summarise such arrangements as a Nash equilibrium.

but that it normally requires a propitious political conjuncture. Thus, Roosevelt was able very successfully to break up US business groups in the wake of the Great Depression. The instruments that were used included limits on the number of levels or tiers, higher taxes on inter-group dividend payments, the effective elimination of consolidated group tax filing, as well as the introduction of constraints on financial institutions acting as controlling shareholders and business groups controlling public utilities.[11] There was broad support for this assault on business groups among the population and elected representatives as they were widely viewed to have behaved abusively, helping to cause and deepen the Great Depression. There was also a consensus that economic concentration needed counteracting. Within a decade, business groups in the United States had largely disappeared.

In a very different context, the US occupation of post-war Japan saw the dismantling of the *zaibatsu* and an attempt at creating dispersed ownership. Although this demarche succeeded in eliminating the previous holding companies and control by family business groups, cross-holdings of shares by corporations and banks subsequently emerged (*keiretsu*).[12] More recently, in the United Kingdom pressure from British institutional investors, concerned by corporate governance problems in business groups, led the government effectively to outlaw them.[13] In 1968, a new Stock Exchange Takeover Rule was introduced which mandated that any acquisition of 30 per cent of a listed company had to be an acquisition of 100 per cent. This was sufficient to prevent pyramids and cross-shareholdings. Similarly, in Israel over the past decade, pressure on business groups has been exerted by limiting the number of levels in pyramidal ownership structures.[14]

7.4.1 Framing Policy Towards Business Groups in Asia

To come to grips with the problems caused by business groups and, hence, the connections world's foundations, several strands of policy

[11] Morck (2005).   [12] McGuire and Dow (2003).   [13] Franks et al. (2004).
[14] Hamdani et al. (2020).

need to be considered simultaneously. We focus on the following main areas: corporate governance, regulations concerning ownership structures (pyramids), mergers and cross-holdings, corporate taxation, inheritance taxes to shift incentives across generation and competition policy.

7.4.1.1 Corporate Governance

Improved corporate governance comprising the leadership, oversight and supervision of boards has the potential to limit the ability of business group dynasties to coordinate their affiliates for the benefit of the family rather than each corporate unit. Corporate governance reforms – such as increased protection for minority shareholders and limits on related party transactions – have been widely promoted and even adopted through Asia. But the impact has been limited, not least because controlling shareholders are powerful and the ability to flout formal rules or codes is widespread.[15] There remains a huge gap between the regulatory framework and actual effectiveness in regulating company behaviour.[16]

Even so, sustained pursuit of certain governance changes could gradually encourage changes in behaviour. These include improved protection of minority shareholders, for example, by giving them legal rights to call shareholder meetings and to limit dilution of shareholdings by majority shareholders. Another important area for improved governance, which could get to the heart of the non-transparent inter-group financial transfers, is to ensure minority shareholders have a say in related party transactions, requiring votes on such transactions at shareholder meetings and veto rights over sales of additional shares. At the same time, requirements for disclosure could be strengthened,

---

[15] Measures, such as use of independent outside directors, have largely been an ineffectual instrument in the Asian context (OECD, 2011).

[16] A recent study found that, 'corporate boards in Asia still lack adequate leadership, supervision and oversight' (Puri, 2020). See also the Asia-Pacific Economic Cooperation (APEC) (2020).

for example requiring publication of plans for related party transactions.[17] Taken together, the combination of such policies might begin to have an impact.

7.4.1.2 Prohibitions and Taxation

Imposing limits on the number of levels permitted in pyramid structures, as also on the number of subsidiaries, is a model that has been widely used to curb the power of business groups. Further options include prohibiting financial and non-financial activities in the same business group. In addition, there are policies designed to restrict cross-holdings, as well as constrain the use of holding companies. South Korea has been the country where a wide variety of attempts using such instruments have been tried since the 1980s. These have included, at various times, a ban on holding companies as well as prohibition of cross-shareholdings. But the latter actually led to a sharp increase in circular shareholding until that in turn was banned. Although business group subsidiaries were subsequently barred from making new internal cross-shareholding investments, the impact has been limited because the law did not affect existing cross-shareholdings. And with respect to holding companies, this policy has over time been reversed. Raising ownership requirements for holding companies – 30 per cent for listed; 50 per cent for unlisted – is now viewed as enabling greater transparency and accountability. These various interventions have all aimed to weaken the incentives for family business groups, yet they have achieved limited results. The resilience and weight of the large South Korean business groups remains almost untouched.

In India, similarly, explicit rules (the Companies Act of 2013) have been put in place for most holding companies to limit the number of tiers and subsidiaries to two, not least to limit subsidiaries being used as a way of hiding debts and siphoning off funds. But, again, they have had limited effect, not just because of the way layers are

---

[17] OECD (2011).

calculated but also because they do not apply to pre-existing subsidiaries. In addition, the agency charged with enforcement – the Ministry of Corporate Affairs – lacks some of the legal and practical ability to do so effectively. The penalties for non-compliance are also relatively trivial. Israel's approach – which also dates to 2013 – has been stricter and more convincing, giving business groups a limited time in which to reduce to no more than two layers while also banning large companies from controlling both financial and non-financial entities.[18]

What these examples show is that prohibitions and more restrictive rules about how businesses can be organised have mostly had limited impact. Business groups are nothing if not inventive in defence of their privileges and many restrictions have been circumvented or largely ignored. That does not mean to say that such policies are misguided: they are not. But they are unlikely by themselves to dent a resilient model. Much the same can be said for the use of discriminatory tax policies. Although in principle higher taxation of dividends could restrict inter-group transfers, most Asian business groups do not rely on dividends for these purposes, using loans and transfer pricing instead. However, this alone does not dilute the broader principle of using taxation to penalise businesses that persist in organising themselves as business groups and hence operating as closely held (and often non-transparent) businesses. Indeed, this can be achieved in part by using the tools of corporate taxation, specifically by imposing additional tax liabilities over and above normal corporate tax rates. Asian economies all have corporate taxation rules in place, although the level of revenues raised varies widely, as does the share in total taxation which ranges between 20 per cent and 40 per cent. In other words, any additional tax liability needs to be superimposed on these current corporate tax rates but only for closely held, family businesses. In general, this involves targeting the business entity instead of individuals. This has the advantage that the

---

[18] Hamdani et al. (2020).

value of any control premium is effectively attributed and taxed to all the relevant members of the family.[19] And it has the additional advantage that we have emphasised in that it targets the format of the business with the aim of reducing the incentive to organise in this way. With respect to the size of the business and its maturity, it is important not to jeopardise entrepreneurial family firms – particularly start-ups – so any such supplementary tax would have to be conditioned on the size of the firm (such as a certain threshold of assets) and, possibly, the time that it had been in existence in that form.

What may also be particularly promising in terms of taxation is the adoption of inheritance or successor taxes. This is indeed the model that the South Koreans have moved towards in the past decade by introducing a 50 per cent inheritance tax rate. As an example, the effective tax rate in 2020 on the value of the deceased chairman of the Samsung Group shares amounted to over 58 per cent, some $10 billion. Although his heirs may settle part of this using cash from dividends, they will likely have to sell property and other assets along with some non-core businesses that have cross-shareholdings in the largest part of the business group, Samsung Electronics. At the same time, the current chairman has already announced that he does not intend to hand over management to his children.[20] More generally, the major business group families view inheritance taxation as a threat to the way their businesses are organised, even if some are likely to use defensive actions such as trusts and other vehicles that limit exposure. Over time, this approach may prove a more effective way of curtailing family business groups than the raft of measures on cross-holdings and other institutional formats that have also been tried. Further, the introduction of inheritance taxation also has the broader advantage of helping address the high levels of inequality that

---

[19] Oh and Zolt (2018) who argue that, for emerging markets, wealth tax add-ons which target and tax particular forms of wealth as part of the existing tax system are preferable to the wholesale introduction of new wealth taxes.

[20] *Financial Times*, 'Samsung's biggest challenge: The Lee family has to reform', 7 February 2021.

have emerged throughout Asia and which we discuss in more detail later in this chapter.

7.4.1.3 Competition Policy

Chapter 4 showed just how much economic concentration has been achieved by the business groups. This signals a major potential role for competition policy. Competition policy operates at the interface of law and business and is implemented through specialised courts or regulatory authorities. It can seek to legislate against firms who are monopolies, for example by setting a legal limit on market share, and then ban any acquisition or merger that takes the firm above such a level. It can also legislate against abuse of monopoly power, without reference to market share, and then judge whether for example pricing is predatory or behaviour towards suppliers or customers is anti-competitive. Key issues in competition law cases must therefore include the definition of the market. However, predatory behaviour such as setting low prices to prevent entry or drive competitors out of the market is hard to prove, especially in business groups where firms can shift costs and overheads around their affiliates. Price fixing, rigging bids for tenders and other cartel activities have also proven very hard to establish.

Because of these factors, judgements under competition law tend to be complex and expensive. They rely for their effectiveness on a specialised legal infrastructure of judges, lawyers and courts, as well as political support in the face of intense business lobbying.[21] Asia – with its well-known deficiencies in terms of institutions and the close nexus of politicians and dynastic business families – therefore faces special difficulties in implementing competition policies.

---

[21] For example, Thomas Philippon (2019) has argued that in recent years the United States has allowed lobbying and campaign contributions to weaken the implementation of competition policy.

At a formal level, the position in Asia with respect to competition policy and associated regulations concerning the control of M&A
and bankruptcy arrangements is that all countries covered in this
book have enacted competition laws, all have bankruptcy laws and
all except Malaysia have merger control laws.[22] Some – as in South
Korea – date back to the 1980s, but most have been adopted within
the last ten to twenty years and some, such as the Philippines and
Thailand only after 2015. However, as Philippon (2019) has pointed
out for the United States, even when the relevant policy structures are
in place, the implementation of competition law may be captured by
the very actors whose behaviour it was enacted to control: large
businesses. The dangers are even more marked in Asia where market
concentration is probably greater, where there are additional complex-
ities resulting from the concentration of overall economic power and
where the connections world is endemic. It is hardly surprising that
there is little evidence that Asian competition authorities are able to
address the complex market power issues posed by business groups.

Indeed, the implementation of competition policy in much of
Asia has been erratic. Take the example of Indonesia where formal
rules seem well designed to address business groups, with prohibitions
against trusts, cross-directorships and majority cross-shareholdings. Yet
there have as yet been no cases against the top four business groups –
Astra, Salim, Lippo or Sinar Mas – that account for nearly 20 per cent of
the country's market capitalisation. The Indonesian Bankruptcy Law
has also been criticised for failing to distinguish between corporate and
individual bankruptcy, as well as inconsistency in its application. In the
Philippines, the competition authority did actually reject an acquisition
bid in 2019[23] but the following year its automatic right to investigate

---

[22] OECD (2018a, b).

[23] Universal Robina Corp's (URC) bid to acquire Roxas Holding's Central Azucarera
Don Pedro Inc. noting that the deal would create a monopoly in the Southern Luzon
sugar market. However, in 2020, URC got approval from the PCC to acquire the
sugar milling and bio-ethanol assets of Roxas Holdings.

M&A was suspended.[24] Similar mixed stories can be told in India and most other democratic countries.

In the autocracies, competition policy, despite bearing many details of design in common, is mostly used to buttress favoured incumbents and/or attain political ends. In China, much of the focus has been on protecting incumbents, including state-owned companies. But the political dimensions are rarely far from the surface. Probably the most glaring example has concerned Alibaba which, as we saw in Chapter 4, has been trying to enter the financial sector through its affiliate, Ant Group. The Group has been offering money management and investment services, including to the dominant state-owned banks. However, the Chinese Competition Commission launched an anti-monopoly investigation into Alibaba in 2020 ordering Ant Group to limit itself to payment services while rectifying 'problems' in areas such as personal lending, insurance and wealth management.[25] Alibaba was at the same time fined for mispricing products while its competitor – Tencent – also began to be investigated in early 2021 for unfair competition on its e-commerce platform – Vipshop – as well as being fined for not declaring past acquisitions. In addition, competition and other government agencies have been used to block entrants into sectors where connected companies exist.

What should we conclude? Asia has largely put in place the formal preconditions for addressing the abuses of business groups, itself an important element in reducing the advantages to their continuation. However, the Asian countries have mostly failed to put anything sufficiently specific and effective in their competition policy to counter the most significant business organisation in their landscape: business groups.

Nevertheless, having a framework for competition policy is an essential step. Perhaps most significantly, it brings the broader issues of the accretion and abuse of economic power into the public domain, providing evidence and data for the press and body politic to consider

---

[24] PCC Press Release (5 October 2020).   [25] *Dow Jones Institutional News* (2020).

and evaluate. Competition policy can act as a brake on the accumulation of further market power, not least if all significant mergers have to be approved by the Competition Office. This may not act to reverse previous accumulations of economic power, but the very threat of the public gaze, let alone expensive court cases, may deter some of the more egregious excesses.

However, to deal with the anti-competitive impact of business groups, the authorities have to move beyond evaluating marginal changes to market structure because of mergers and acquisitions towards making judgements about existing market structure and the impact of overall concentration in the economy. Drawing on Israel's recent experience, it would be advisable if policy took into account overall levels of market concentration, as well as market concentration in particular sectors, when assessing market power and abuse of competition.

A simple way to approach this is to set limits to the maximum market share that a firm or business group can hold – for example, to 30 per cent – and then to require existing firms to be broken up to ensure the thresholds are not exceeded, sector by sector. The thresholds might be lowered for cases when firms are affiliates of business groups, so as to counter the group's ability to leverage resources and market power through their position in other markets. Thus, for example, firms with market shares in excess of 30 per cent would be required to sell or liquidate capacity until the threshold was met. Firms who were business group affiliates with, say, more than a 5 per cent share of the overall economy might instead face a threshold of 25 per cent. Although this sort of approach would be more stringent than in most advanced economies, the starting point in Asia is one where there is far more market power vested in a limited number of business groups. Hence, a much tougher policy stance will be required to level conditions facing incumbents with that for new entrants as well as preventing abuses.

7.5 BRINGING TRANSPARENCY

This brings us to the other side of the connections world: that of the politicians. As we have argued throughout this book, politicians draw much benefit – both for themselves, sometimes for their constituents and often for their political parties – from connections. Realistically, this is not something from which most politicians will be willing to recuse themselves in the absence of effective sanctions and associated loss of reputation. The hope that a free and curious press – as exists in India – would shine sufficient light on murky relationships and deals has often been satisfied but, sadly, has rarely proven by itself to be a sufficient source of discipline. In many other countries, notably the autocracies – not even that light has been shone. What this obviously suggests is that greater transparency through the media, whilst inherently desirable, will rarely, if ever, be sufficient to rein in politicians' self-seeking behaviour.

There are, however, a host of other, supportive measures that over time might be expected to erode the worst excesses and to cause some modification in behaviour. Rules regarding political contributions need tightening. For example, with the exception of South Korea, no other country has a ban on contributions by companies to political parties. In fact, in most places even companies that have partial state ownership can make contributions. Likewise, there are few, if any, restrictions on the amount that can be donated to politicians whether by individuals or corporates.

There is also the use of asset and interest declarations. Most countries, other than Bangladesh, do notionally mandate such declarations. But much depends on how such reporting is designed let alone implemented. There are, in fact, a variety of possible formats. For example, in the United Kingdom, the main instrument is the mandated, periodic reporting by all parliamentarians through a register of interests. Ministers have to declare any financial interests that are relevant to their responsibilities, and which could conflict with their duty to the public. By the same criteria, they also need to disclose the interests of their close family which extends to siblings, parents, spouse and in-laws. Parliamentarians who are not ministers have lower degrees of responsibility for reporting financial interests. Other countries have adopted different formats, but the main issues always concern the extent of coverage, frequency of reporting and the extent to which the information that is compiled is made available. Clearly, electronic reporting has many practical advantages and seems to have been associated with greater transparency and public accountability.[26] Most decentralised systems of reporting that involve agencies verifying the filings of their own officials end up being compromised.

The broad point is that pushing for greater automatic disclosure along with easy and wide access to that information is a crucial prerequisite. That needs to be accompanied by parallel measures aimed at limiting corruption in public spending decisions, including contract assignment. Such measures include frequent and external audits. At the same time, some countries have achieved traction by setting up dedicated anti-corruption agencies – in Indonesia's case, KPK – that are tasked with pursuing corrupt practices and have sufficient autonomy and political support to take on sensitive cases. As with measures aimed at reining in business groups, however, experience strongly suggests that the effects are rarely rapid. What they can provide is a framework for improvement with associated instruments that can be mobilised in the future.

7.6 OTHER BIG PRESSURE POINTS

The forward-looking challenges of innovation, productivity and further income growth are compounded by a set of associated forces that are bearing down on Asia. Some are specific to the connections world, but others are a function of wider and highly disruptive changes to technology whose impact is being increasingly felt in most parts of

---

[26] World Bank (2020c).

the global economy. Specifically, the jobs challenge stands as a major rebuke to the way in which connections-driven growth has worked out. That challenge is put into yet sharper relief by the growing application of labour-substituting technologies. In addition, the connections world will be increasingly pinched by the extraordinary inequalities that it has generated. The consequences of those inequalities are likely to include a higher proclivity for distributional disputes spilling into political upheaval. This will potentially be most consequential in regimes where reconciliation of political differences is, almost by design, highly difficult, if not impossible.

7.6.1 The Jobs Challenge

We saw in Chapter 6 how the jobs challenge stalks Asia's economies and how the way that the connections world functions increasingly partitions the economy into its privileged parts – the formal jobs that connected companies create – and the far larger parts that lack those privileges and hence are consigned to informality and low productivity. Moreover, this situation with regard to employment does not exist in aspic. There is the looming matter of how the Asian economies will adapt to changes in the wider landscape of work driven by recent waves of technological change. These shifts – as is the case more widely – have serious potential to disrupt how products are made or services generated. Specifically, they will potentially affect not just the demand for labour in aggregate, but also its composition. Indeed, rapid technological change involving machine and AI is already in process. The deployment of robots – one aspect of this change – is already advancing in factories in Asia as much as in the rich world. South Korea has the most industrial robots in use relative to the number of employees in the world.[27]

The possible consequences of these changes are still unclear. Some (not very scientific) estimates attach very high probabilities of employment contraction with up to half of occupations in China and

---

[27] OECD (2017).

India being at serious risk.[28] But measures of technological risk for specific tasks, rather than occupations, paint a less stark and far more nuanced picture. Although labour costs in Asia still remain relatively low, they have been rising quite sharply – notably in China. And this may increase the attraction to employers of substituting capital for labour, particularly if allied regulatory constraints are seen as a serious impediment.

Technological advances are certainly changing how goods and services are created in Asia and elsewhere, including at the factory-floor level. The current phase of technological change combines hardware and software through connectivity. This permits greater flexibility and effectiveness in the execution of tasks. As such, networked systems with in-built connections allow continuous interaction with the physical world, not least through the exploitation of big data. Firms applying recent generation technologies in production also tend to have far higher engagement with information-intensive business services.

A few Asian economies – notably China, South Korea and Taiwan – already rank among the top countries for advanced digital production, as measured by the number of patents. India and Singapore are actively engaging as producers while Thailand, Malaysia and Vietnam have reasonably high values of imports.[29] Although most Asian economies have a mix of – mostly earlier vintage – technologies in play with very different levels of sophistication and automation, the contribution of recent and current technology vintages is growing. For example, in Thailand they have been adopted by around a quarter of firms. Indeed, about 15 per cent of the stock of industrial robots is in emerging economies and the greatest part of that is in Asia. The scope for further adoption is considerable if the main constraints – notably a lack of the right skills or capabilities – are addressed.

---

[28] See, for example, McKinsey (2017).   [29] See UNIDO (2020).

There is also the matter of Asian economies' insertion in global value or supply chains (GVCs). GVCs grew rapidly from the 1990s but since 2010 this has been partly reversed because of their perceived complexity and fragility, further highlighted by the COVID-19 pandemic. Perhaps not surprisingly, the ways in which production is carried out in GVCs tends to conform to routines and norms that exist for the industry or activity and which are effectively set without primary reliance on the relative labour cost advantage of the emerging market producer.[30] The automobile industry is a good example. More intensive use of robots (the industry accounts for over two-fifths of all robot use globally) along with more skilled labour has increasingly become the norm.

New technologies carry wider implications. One is that they will permit or accelerate reorganisation of supply chains, including moving production back to the advanced economies.[31] Aspirations by Asian companies to capture more highly skilled work has also been compromised by the continuing locational advantages that rich countries possess, notably the powerful combination of transparent business rules and practices along with the presence of large numbers of skilled professionals.[32] Falling relative capital costs will also promote relocation of those activities, where proximity to core markets and demand is an important consideration.[33]

In short, Asia already had a struggle on its hands to create jobs for its citizens. Those jobs that have been created – and they are numerous – have mostly been low productivity and low skill. The connections world has proven limited in its ability to create large numbers of good jobs. Now, those tensions are being exacerbated by the scale and scope of technological advance which also challenges some assumptions drawn from previous experience. With the ICT revolution, for example, although some jobs were destroyed, many

---

[30] UNIDO (2020) found that in Thailand and Vietnam participation in a GVC raised the probability of adopting advanced technology.
[31] Rodrik (2018).   [32] As argued by Iversen and Soskice (2019).
[33] Dachs and Seric (2019).

new ones were created. Technology complemented labour as much, if not more, than it substituted. But with AI and new generation technologies, this assumption may not be valid. Indeed, some have argued that space has been created for technology to develop in ways that are no longer limited by the need for human emulation.[34] To the extent that this is the case, Asia, as much as the rich world, may face a future with less work and certainly fewer good jobs, thereby worsening yet further the existing situation. Intriguingly, some governments – notably China – have alighted on AI as an area of strategic importance and as a potential answer to rising wages. For a government that still promotes its socialist credentials, this engagement may, however, prove to be truly bittersweet, insofar as it grossly aggravates the jobs challenge.

7.6.2 Inequality and the Spoils of Connections

Asia's ascent has been accompanied by rapid increases in inequality in both income and wealth. Four decades ago, South Asian income inequality was relatively low, but it has mostly increased since then. In Southeast Asia, inequality was already at a higher level than South Asia – sometimes, in Malaysia, Philippines and Thailand significantly higher – and although it has declined somewhat in subsequent decades, it has still left these countries with substantial inequality. In China and Vietnam, inequality in the 1980s was also already higher than in South Asia and has since drifted upwards. Only in the richest economy – South Korea – has inequality remained broadly stable and relatively low over recent decades.

There is substantial evidence that more unequal countries tend to under-perform over the longer term.[35] In addition, economies with high inequality tend to be susceptible to upheaval, particularly if that inequality runs alongside a political regime that has limited capacity for adaptation or reform. By definition, autocracies tend to be significantly more rigid and, hence, more prone to blowing up. In such

---

[34] For example, Susskind (2020).   [35] Milanovic (2016).

systems, distributional contests are difficult to resolve because they lack the mechanisms and institutions that can engineer changes, particularly when it involves existing beneficiaries having to sacrifice.

How has the connections world's salience affected the distributional paths of income and wealth? Our book has argued that there have been a series of important and long-lasting consequences. Income inequality has grown as a result of the labour market's dualism and the large disparities in relative earnings between formal and informal sectors. Further, the tendency to concentration and market power, documented in Chapter 4, also shows how some companies, business groups and families have been able to reap outsized benefits from Asia's growth spurt.

The rapid growth in wealth has once again cut across political systems. In all of the countries covered in this book, not only have average wealth levels soared but so has wealth inequality. In Asia – with the exception of South Korea – the Gini measure of wealth inequality now mostly matches, and sometimes exceeds, that in the rich world. In short, as the wealth of these nations has ballooned, much of it has come to be concentrated in relatively few hands at the top of the wealth pyramid.

Nothing illustrates this better than the examples of both China and India despite being archetypically different political systems. In China, although it was hard to know what share of wealth was held by particular groups pre-1980, it has been estimated that the top 1 per cent then held roughly 15 per cent of net personal wealth, just about equivalent to the share of the bottom 50 per cent. By 2000 the share of the top 1 per cent had shifted to 20 per cent before jumping to over 30 per cent in 2019.[36] The path has followed a similar direction and, perhaps even steeper, ascent in India. There, the share of the top 1 per cent went from around 10 per cent in 1978 to 23 per cent in 2000 and around 40 per cent in 2019. By this latter date, in both countries, the bottom 50 per cent controlled around 5 per cent of net wealth. Placed

---

[36] Credit Suisse (2019) and World Inequality Database (2020).

in wider context, the combined total wealth of these countries amounted to over 21 per cent of total global wealth (as against a mere 4.3 per cent in 2000), with their top 1 per cent controlling over 7 per cent of global wealth. Expressed differently, in these two countries with their combined populations of 2.75 billion – a quarter of human-kind – a grand total of 928 individuals had wealth of over $500 million.[37]

Needless to say, when peering inside that slim tranche of the top 1 per cent, what we find are a mix of individuals and families who mostly own and control major businesses, commonly in the form of business groups. Some of that wealth has been inherited, as many of these family businesses have been around for considerable amounts of time, but some of it has been newly made, most particularly in China. In short, the connections world has spawned a tremendous growth in wealth at the top, a concentration that is only likely to be perpetuated unless the barriers to competition and rivalry are eroded.

The extraordinarily sharp rise in wealth and its concentration begs the obvious question of whether there need to be countervailing policies and if so, of what design. In particular – and in common with many rich countries – there have been calls to tax wealth. Such taxation could focus, in principle, on both financial and non-financial assets. It could include taxes on the value of assets or revenues as well as, potentially, the application of property taxes. Property is a major component of Asian wealth and a progressive property tax with sens-ible thresholds would be a potentially powerful instrument. In add-ition, the transfer of wealth across generations could be addressed by introducing or extending inheritance taxes: an issue that we have discussed specifically in the context of business groups.

Critics of possible redistribution argue that premature attempts at forcing greater equality will throttle the sources of Asia's dyna-mism and send the wrong signals. Moreover, some of the forces that have been pushing inequality upwards are likely to attenuate, as

---

[37] Credit Suisse (2019).

factors such as increased education and skills, as well as the broader demographics, kick in and improve the relative prospects of wage earners, as the path of Chinese formal sector wages may already indicate.

For sure, evidence from other historical episodes of rapid growth points to often sizeable increases in inequality and extravagant winners. Subsequently, the excesses of these various capitalist gilded ages have been mollified, whether through regulatory actions – think of Standard Oil – or through the introduction of more progressive taxation. And, in the lens of Asia's recent past, returning to the sorts of confiscatory policies and high marginal tax regimes of earlier decades would probably yield limited revenues, accelerating even more massive movement of capital offshore.

There are additional reasons for why simply focusing on redistributive measures risks misfiring, let alone being ineffectual. For a start, most Asian economies still have relatively low tax bases, raising between 8 per cent and 20 per cent of GDP in taxes. Most of that is raised through indirect taxes – such as VAT – with both personal and corporate taxation being held down by not only limited coverage but, very importantly, limited institutional capacity. For instance, most of the informal economy pays neither personal income tax nor corporate taxes and even those individuals or entities that are clearly liable are often able to avoid or dilute their exposure. In a recent year, only a quarter of Pakistan's elected parliamentarians had filed tax returns and many other well-to-do individuals were similarly not registered with the tax authorities.[38] Such stories abound across all the countries and underline the point that taxation possibilities are radically bounded by conditions on the ground.

As to the taxation of wealth in Asia, the picture there is even more restrictive with few attempts to date and mostly paltry results. Only a very few places – Philippines, Thailand and Vietnam – have introduced inheritance taxes, although – as noted earlier – South

[38] ESCAP (2017).

Korea is an outlier with an inheritance tax with one of the highest basic rates (50 per cent). More generally, only China has managed to collect more than 0.5 per cent of GDP through taxation of wealth and that has mainly been through property taxation. Putting this in context, the OECD economies raise around 2 per cent of GDP on average through such taxes, mostly targeted at inheritance. Only three OECD economies currently impose explicit wealth taxes not conditioned on inheritance.

But simply letting the status quo stand also carries serious risks, not least with respect to the consequences of the connections world for Asia's future. We have already made the case for improving transparency and oversight along with measures that tackle head on the accumulations of market power and the associated attenuation of rivalry that are central features of the connections world. In short, instead of focusing mainly on redistributive measures, a better approach would be to start addressing the fundamental reasons for why the connections world continues to thrive and, hence, the high levels of inequality that it spawns.

Over and above these measures, there may also be grounds for concentrating on the uber-wealthy. However, as we have already seen, their numbers are rather small and their capacity to use complex financial instruments, including offshore vehicles, is considerable. This will limit the yield from taxes aimed specifically at this group. For example, it has been estimated that imposing a 0.5 per cent wealth tax on the 0.3 per cent of the Chinese population with assets above $1 million would raise less than $70 billion: no more than 3 per cent of total tax revenues. The advantage of targeting business entities using adaptations of existing corporate tax instruments is that these have a specific objective of switching incentives for family-based business groups. To the extent that they are successful and some of the attractions of the connections world pale and institutions improve making meaningful taxation of capital possible, these adaptations could give way to more generalised taxation of wealth.

7.7   SOME FINAL CONSIDERATIONS

How should we then view Asia's potential for a further ascent of the income and development mountains? And, perhaps most importantly, how sustainable will be the strategies and features that have already taken it so far? We have argued that the further ascent is compromised by the connections world that spans the region and which, as have repeatedly seen, has proven so dexterous and effective in entrenching themselves.

There are two additional questions we should raise. The first concerns whether what exists in Asia is fundamentally different to what existed in the rich countries when they were at comparable levels of income and development. The second concerns what impact the connections world will have on the big global questions of our age, notably climate change, biodiversity and, hence, sustainability.

The idea that institutional and other forms that have aided rapid development ultimately need to be superseded is not a new one. After all, mercantilism, nineteenth-century laissez-faire capitalism and, later, central planning have all faded away as they moved from being supportive for growth to being constraints. The Asian connections world is distinguished from these experiences not least due to the corrosive, but intricate, intertwining of business and political interests and the massive entrenchment that has consequently occurred. The forces driving these economies towards more open and more competitive arrangements remain unequally poised. In addition, this entrenchment draws on something that is hugely potent, the wider cultural features that support and sustain reliance on family business. With these traits also comes an absence of traditions and practices that favour reliance on arms-length transacting. These attributes are compounded by the way in which the public sector – notably the state-owned banks – has proved integral to maintaining flows of finance on terms and conditions that are neither arms-length nor, in many instances, warranted. Putting all these features together, it seems highly likely that incremental change may struggle to break through and, as a consequence, it may require more radical measures to build sufficient momentum to overcome the current organisational inertia.

Although many of Asia's challenges can be traced to the connections world, it is clear that many also bear strong resemblance to those actually facing the rich world. After all, much of informed public debate in the latter is fixated on inadequate productivity growth, the threat to jobs from technology and AI, the declining vitality of competitive forces, alongside rising inequalities in income and wealth. Evidently, these challenges are driven by many of the same factors as those facing Asia. Even so, the rich world still has some important advantages. These include the ability to draw on the depth of its skill base, social capital and infrastructure, along with well-established and supportive institutional arrangements, such as the rule of law and enforcement of property rights, and, perhaps most importantly, relatively robust political institutions. These attributes are largely absent in Asia and consequently make future adaptation more challenging.

We have also not yet reflected directly on the paramount policy question of our time: climate change and the wider matter of sustainability, both for individual countries and for the planet as a whole. Undoubtedly, the relentless pursuit of growth by the Asian economies has exacerbated global warming. In 1990, India generated 500 million tons of CO_2 emissions and China some 2,400 million tons. This compared to >5,000 million tons in the United States. By 2020, India's emissions approached 2,600 million tons, nearly 7 per cent of the global total, while China's had risen more than fourfold to >11,500 million tons or 30 per cent of the total. The United States was more or less constant at 5,100 million tons or 13 per cent of the total.[39] While the EU, the United States and Japan have all been reducing emissions, they continue to rise rapidly across Asia, especially in China and India. The Chinese have announced that they are

[39] EDGAR (2020).

aiming for emissions to peak in 2030 with carbon neutrality by 2060. Other Asian countries have been far less ambitious. And it should be noted that the projected rise in emissions, including from coal, in the coming decades can mainly be traced to both India and China. This may be unsurprising given that much of the rapid Asian growth of recent decades has been fuelled by coal-fired electricity generation, as well as the broader reliance on industries that are carbon-intensive. Needless to say, in Asia natural resources and the energy sector have continued to be one of the heartlands of the connections world. The lobbying power of these interests remains formidable and, although faced with a growing challenge from renewables, seems unlikely to fold. What this suggests, of course, is that the ability, or otherwise, of Asia to mutate its form of capitalism away from the connections world will have direct ramifications for all of us, wherever we may live.

References

Acemoglu, D. and Robinson, J. A. (2012) *Why Nations Fail: The Origins of Power, Prosperity, and Poverty*. 1st ed. New York: Crown.

ADB (2012) *Asian Development Outlook 2012: Confronting Rising Inequality*. Manila: ADB.

Ahrens, N. (2013) *China's Competitiveness: Huawei*. Washington, DC: Center for Strategic & International Studies.

Akcigit, U., Baslandze, S. and Lotti, F. (2018) *Connecting to Power: Political Connections, Innovation, and Firm Dynamics*. NBER Working Paper No. 25136. Cambridge, MA: National Bureau of Economic Research.

Almeida, H., Kim, C. and Kim, H. B. (2015) 'Internal Capital Markets in Business Groups: Evidence from the Asian Financial Crisis', *The Journal of Finance*, 70 (6), pp. 2539–86.

Almeida, H., Park, S. Y., Subrahmanyam, M. G. and Wolfenzon, D. (2011) 'The Structure and Formation of Business Groups: Evidence from Korean Chaebols', *Journal of Financial Economics*, 99(2), pp. 447–75.

Amsden, A. H. (1992) *Asia's Next Giant: South Korea and Late Industrialization*. Rev. ed. New York/Oxford: Oxford University Press.

Amsden, A. H. (2009) *Escape from Empire: The Developing World's Journey through Heaven and Hell*. Cambridge, MA: MIT Press.

Anderson, R. C. and Reeb, D. M. (2003) 'Founding-Family Ownership and Firm Performance: Evidence from the S&P 500', *The Journal of Finance*, 58(3), pp. 1301–28.

Anjumol, K. S., Sandeep, S., Rajan, N. and Neethu, N. (2019) *A Study Report on Reliance Jio Infocomm Limited*. Thiruvananthapuram, India: Indian Institute of Space Science and Technology.

APEC (2020) *Protecting Minority Investors in Privately Held Companies in APEC*. Singapore: APEC; USAID.

Appelbaum, R. P., Cao, C., Han, X., Parker, R. and Simon, D. (2018) *Innovation in China: Challenging the Global Science and Technology System*. 1st ed. Cambridge: Polity Press.

Asian Development Bank (2011) 'Asia 2050: Realizing the Asian Century'. Manila, Philippines. Available at: www.adb.org/publications/asia-2050-realizing-asian-century

Audretsch, D. B. (1995) *Innovation and Industry Evolution*. Cambridge, MA: MIT Press.

Audretsch, D. B. and Feldman, M. P. (1996) 'R&D Spillovers and the Geography of Innovation and Production', *The American Economic Review*, 86(3), pp. 630–40.

Baek, J.-S., Kang, J.-K. and Lee, I. (2006) 'Business Groups and Tunneling: Evidence from Private Securities Offerings by Korean Chaebols', *The Journal of Finance*, 61(5), pp. 2415–49.

Baily, M., Zitzewitz, E., Bosworth, B. and Westphal, L. (1998) 'Extending the East Asian Miracle: Microeconomic Evidence from Korea', *Brookings Papers on Economic Activity. Microeconomics*, 29, pp. 249–321.

Baldwin, R. E. (2016) *The Great Convergence: Information Technology and the New Globalization*. Cambridge, MA: The Belknap Press of Harvard University Press.

BAPPENAS (2018) 'Indonesia Vision 2045'. Jakarta: Ministry of National Development Planning.

Bartelsman, E., Haltiwanger, J. and Scarpetta, S. (2004) *Microeconomic Evidence of Creative Destruction in Industrial and Developing Countries*. IZA Discussion Paper No. 1374. Bonn: Institute of Labor Economics (IZA).

Baten, J. (ed.) (2016) *A History of the Global Economy: 1500 to the Present*. Cambridge: Cambridge University Press.

Beasley, W. G. (2018) *The Meiji Restoration*. Stanford, CA: Stanford University Press.

Belenzon, S. and Berkovitz, T. (2010) 'Innovation in Business Groups', *Management Science*, 56(3), pp. 519–35.

Bertrand, M., Mehta, P. and Mullainathan, S. (2002) 'Ferreting Out Tunneling: An Application to Indian Business Groups', *The Quarterly Journal of Economics*, 117(1), pp. 121–48.

Bertrand, M. and Schoar, A. (2006) 'The Role of Family in Family Firms', *Journal of Economic Perspectives*, 20(2), pp. 73–96.

Bhaumik, S. K., Estrin, S. and Mickiewicz, T. (2017) 'Ownership Identity, Strategy and Performance: Business Group Affiliates versus Independent Firms in India', *Asia Pacific Journal of Management*, 34(2), pp. 281–311.

Bloomberg (2020) 'Vast Numbers of Unemployed Will Undermine China's Recovery'. *Bloomberg*, 13 May. Available at: www.bloomberg.com/news/articles/2020-05-13/vast-numbers-of-unemployed-will-undermine-china-s-recovery

Bolt, J. and van Zanden, J. L. (2020) 'Maddison Project Database 2020'. Maddison Project Database. Available at: www.rug.nl/ggdc/historicaldevelopment/maddison/releases/maddison-project-database-2020

Bosworth, B. and Collins, S. M. (2008) 'Accounting for Growth: Comparing China and India', *Journal of Economic Perspectives*, 22(1), pp. 45–66.

Bowen, H. P. and Wiersema, M. F. (2005) 'Foreign-Based Competition and Corporate Diversification Strategy', *Strategic Management Journal*, 26(12), pp. 1153–71.

Braudel, F. (1992) *Civilization and Capitalism, 15th–18th Century*. Berkeley, CA: University of California Press.

Business Standard (2020) *Infosys Company History*. Available at: www.businessstandard.com/company/infosys-2806/information/company-history

Bussolo, M., de Nicola, F., Panizza, U. and Varghese, R. (2019) *Political Connections and Financial Constraints: Evidence from Transition Countries*. Policy Research Working Paper No. 8956. Washington, DC: World Bank.

Calvino, F., Criscuolo, C. and Menoni, C. (2015) *Cross-country Evidence on Start-up Dynamics*. OECD Science, Technology and Industry Working Paper No. 2015/06. Paris: OECD Publishing.

Carney, M. (2008) 'The Many Futures of Asian Business Groups', *Asia Pacific Journal of Management*, 25(4), pp. 595–613.

Carney, M. and Gedajlovic, E. (2002) 'The Co-evolution of Institutional Environments and Organizational Strategies: The Rise of Family Business Groups in the ASEAN Region', *Organization Studies*, 23(1), pp. 1–29.

Carney, M., Gedajlovic, E., Heugens, P., Essen, M. and Oosterhout, J. (2011) 'Business Group Affiliation, Performance, Context, and Strategy: A Meta-Analysis', *Academy of Management Journal*, 54(3), pp. 437–60.

Carney, M., van Essen, M., Estrin, S. and Shapiro, D. (2017) 'Business Group Prevalence and Impact across Countries and Over Time: What Can We Learn from the Literature?', *Multinational Business Review*, 25(1), pp. 52–76.

Carney, M., van Essen, M., Estrin, S. and Shapiro, D. (2018) 'Business Groups Reconsidered: Beyond Paragons and Parasites', *Academy of Management Perspectives*, 32(4), pp. 493–516.

Casanova, L. and Miroux, A. (2018) *Emerging Markets Multinationals Report 2018*. Ithaca, NY: Emerging Markets Institute, Cornell S.C. Johnson College of Business, Cornell University.

Castellacci, F. (2015) 'Institutional Voids or Organizational Resilience? Business Groups, Innovation, and Market Development in Latin America', *World Development*, 70, pp. 43–58.

CB Insights (2020) *The Complete List of Unicorn Companies, CB Insights*. Available at: instapage.cbinsights.com/research-unicorn-companies

Cestone, G., Fumagalli, C., Kramarz, F. and Pica, G. (2016) *Insurance Between Firms: The Role of Internal Labor Markets*. No. 11336. London: Centre for Economic Policy Research (CEPR).

Chambers, D. S. (1970) *The Imperial Age of Venice, 1380–1580*. New York: Harcourt Brace Jovanovich Publishers.

Chang, S. J. and Choi, U. (1988) 'Strategy, Structure and Performance of Korean Business Groups: A Transactions Cost Approach', *The Journal of Industrial Economics*, 37(2), pp. 141–58.

Chari, M. D. R. (2013) 'Business Groups and Foreign Direct Investments by Developing Country Firms: An Empirical Test in India', *Journal of World Business*, 48(3), pp. 349–59.

Charmes, J. (2012) 'The Informal Economy Worldwide: Trends and Characteristics', *Margin: The Journal of Applied Economic Research*, 6(2), pp. 103–32.

Chen, K.-C. and Redding, G. (2017) 'Collaboration and Opportunism as a Duality Within Social Capital: A Regional Ethnic Chinese Case Study', *Asia Pacific Business Review*, 23(2), pp. 243–63.

Chen, T. and Kung, J. K. (2019) 'Busting the "Princelings": The Campaign Against Corruption in China's Primary Land Market', *The Quarterly Journal of Economics*, 134(1), pp. 185–226.

Chen, T.-J. and Ku, Y.-H. (2016) 'Rent Seeking and Entrepreneurship: Internet Startups in China', *Cato Journal*, 36(3), pp. 659–89.

Cheong, K.-C., Lee, P.-P. and Lee, K.-H. (2015) 'The Internationalisation of Family Firms: Case Histories of Two Chinese Overseas Family Firms', *Business History*, 57(6), pp. 1–21.

Cheung, Y.-L., Rau, R. and Stouraitis, A. (2006) 'Tunneling, Propping, and Expropriation: Evidence from Connected Party Transactions in Hong Kong', *Journal of Financial Economics*, 82(2), pp. 343–86.

Chhibber, A. (2018) *India's Public Sector Enterprises: Why the Business of Government is Not Business*. New Delhi: National Institute of Public Finance and Policy and the Federation of Indian Chamber of Commerce and Industry.

Choi, J. J., Jo, H., Kim, J. and Kim, M. (2018) 'Business Groups and Corporate Social Responsibility', *Journal of Business Ethics*, 153(4), pp. 931–54.

Choi, P. P. (2016) 'Evolution of Samsung Group and Its Central Office: Imperfect Market and Capacity-Building', *Asian Business & Management*, 15(5), pp. 370–98.

Christensen, C. M., Raynor, M. E. and McDonald, R. (2015) 'What Is Disruptive Innovation?', *Harvard Business Review*, 1 December, pp. 44–53.

Claessens, S., Djankov, S. and Lang, L. H. P. (2000) 'The Separation of Ownership and Control in East Asian Corporations', *Journal of Financial Economics*, 58(1), pp. 81–112.

Claessens, S., Fan, J. P. H. and Lang, L. H. P. (2006) 'The Benefits and Costs of Group Affiliation: Evidence from East Asia', *Emerging Markets Review*, 7(1), pp. 1–26.

Cohen, W. M. and Levinthal, D. A. (1990) 'Absorptive Capacity: A New Perspective on Learning and Innovation', *Administrative Science Quarterly*, 35(1), pp. 128–52.

Colli, A. and Colpan, A. M. (2016) 'Business Groups and Corporate Governance: Review, Synthesis, and Extension', *Corporate Governance: An International Review*, 24(3), pp. 274–302. DOI: http://doi.org/10.1111/corg.12144

Colpan, A. M., Hikino, T. and Lincoln, J. R. (eds) (2010) *The Oxford Handbook of Business Groups, The Oxford Handbook of Business Groups*. Oxford, UK: Oxford University Press.

Commander, S. (ed.) (2005) *The Software Industry in Emerging Markets*. Cheltenham, UK/Northampton, MA: Edward Elgar Publishing.

Commander, S., Kangasniemi, M. and Winters, L. A. (2004) 'The Brain Drain: Curse or Boon? A Survey of the Literature', in Baldwin, R. E. and Winters, L. A. (eds) *Challenges to Globalization: Analyzing the Economics*. Chicago: University of Chicago Press, pp. 235–278.

Commander, S. and Poupakis, S. (2020) *Political Networks Across the Globe*. IZA Discussion Paper 13103. Bonn: Institute of Labor Economics (IZA).

Coppedge, M., Gerring, J., Knutsen, C. H., Lindberg, S. I., Teorell, J., Altman, D., Bernhard, M., Fish, M. S., Glynn, A., Hicken, A., Luhrmann, A., Marquardt, K. L., McMann, K., Paxton, P., Pemstein, D., Seim, B., Sigman, R., Skaaning, S-E., Staton, J., Wilson, S., Cornell, A., Alizada, N., Gastaldi, L., Gjerløw, H., Hindle, G., Ilchenko, N., Maxwell, L., Mechkova, V., Medzihorsky, J., von Römer, J., Sundström, A., Tzelgov, E., Wang, Y., Wig, T. and Ziblatt, D. (2020). *V-Dem [Country–Year/Country–Date] Dataset v10*. Varieties of Democracy (V-Dem) Project. DOI: https://doi.org/10.23696/vdemds20

Crabtree, J. (2018) *The Billionaire Raj: A Journey Through India's New Gilded Age*. London, UK: One World.

Credit Suisse (2019) *Global Wealth Databook 2019*. Zurich: Credit Suisse Research Institute.

Crossan, M. M. and Apaydin, M. (2010) 'A Multi Dimensional Framework of Organizational Innovation: A Systematic Review of the Literature', *Journal of Management Studies*, 47(6), pp. 1154–91.

Cruz, C., Labonne, J. and Querubín, P. (2017) 'Politician Family Networks and Electoral Outcomes: Evidence from the Philippines', *American Economic Review*, 107(10), pp. 3006–37.

Dachs, B. and Seric, A. (2019) *Industry 4.0 and the Changing Topography of Global Value Chains*. Department of Policy Research and Statistics Working Paper 10/2019. Vienna: UNIDO.

Dalrymple, W. (2019) *The Anarchy: The Relentless Rise of the East India Company*. New York: Bloomsbury.

Davidsson, P. (2016) *Researching Entrepreneurship: Conceptualization and Design*. 2nd ed. New York: Springer.

De Backer, K. and Miroudot, S. (2013) *Mapping Global Value Chains*. OECD Trade Policy Paper No. 159. Paris: OECD Publishing.

De Soto, H. (1989) *The Other Path: The Invisible Revolution in the Third World*. 1st ed. New York: Harper & Row.

Dieleman, M. and Sachs, W. (2008) 'Coevolution of Institutions and Corporations in Emerging Economies: How the Salim Group Morphed into an Institution of Suharto's Crony Regime', *Journal of Management Studies*, 45(7), pp. 1274–300.

Dow Jones Institutional News (2020) 'China Tells Ant to Refocus on Payments Business', 27 December. Available at: www.morningstar.com/news/dow-jones/20201227497/china-tells-ant-to-refocus-on-payments-business

Eichengreen, B., Park, D. and Shin, K. (2013) *Growth Slowdowns Redux: New Evidence on the Middle-Income Trap*. NBER Working Paper No. 18673. Cambridge, MA: National Bureau of Economic Research. DOI: http://doi.org/10.3386/w18673

EDGAR (2020) 'CO_2 Emissions for All World Countries from 1970 to 2019'. Available at: edgar.jrc.ec.europa.eu/overview.php?v=booklet2020

Equitymaster (2020) *Indian Software Industry Report*. Available at: www.equitymaster.com/research-it/sector-info/software/Software-Sector-Analysis-Report.asp

ESCAP (2017) *Prospects for Progressive Tax Reform in Asia and the Pacific*. Bangkok: ESCAP.

Estrin, S., Hanousek, J., Kocenda, E. and Svejnar, J. (2009) 'The Effects of Privatization and Ownership in Transition Economies', *Journal of Economic Literature*, 47(3), pp. 699–728.

Estrin, S., Mickiewicz, T., Stephan, U. and Wright, M. (2019) 'Entrepreneurship in Emerging Markets', in Grosse, R. E. and Meyer, K. (eds) *The Oxford Handbook of Management in Emerging Markets*. Oxford, UK: Oxford University Press.

Estrin, S., Cote, C., Shapiro, D. and Nunner, K. (2020) *Daimler: Motoring at the Speed of China*. Ivey Case No. 9B20M023.

Estrin, S. and Pelletier, A. (2018) 'Privatization in Developing Countries: What Are the Lessons of Recent Experience?', *The World Bank Research Observer*, 33(1), pp. 65–102.

Estrin, S., Poukliakova, S. and Shapiro, D. (2009) 'The Performance Effects of Business Groups in Russia', *Journal of Management Studies*, 46, pp. 393–420.

Estrin, S. and Prevezer, M. (2011) 'The Role of Informal Institutions in Corporate Governance: Brazil, Russia, India, and China Compared', *Asia Pacific Journal of Management*, 28(1), pp. 41–67.

Fagerberg, J., Mowery, D. C. and Nelson, R. R. (eds) (2005) *The Oxford Handbook of Innovation*. New York: Oxford University Press.

Financial Times (2020) 'Modi's Rockefeller: Gautam Adani and the Concentration of Power in India', 13 November. Available at: www.ft.com/content/474706d6-1243-4f1e-b365-891d4c5d528b

Forbes World's Billionaires List 2020 (2020). Available at: www.forbes.com/worlds-billionaires/?sh=10de2b295864

Fortune (2016) *Global 500*. Available at: fortune.com/global500/

Franks, J. R., Mayer, C. and Rossi, S. (2004) *Spending Less Time with the Family: The Decline of Family Ownership in the UK. SSRN Scholarly Paper ID 493504.* Rochester, NY: Social Science Research Network.

Freeman, C. (1995) 'The "National System of Innovation" in Historical Perspective', *Cambridge Journal of Economics*, 19(1), pp. 5–24.

Freund, C. L. (2016) *Rich People Poor Countries: The Rise of Emerging-Market Tycoons and Their Mega Firms*. Washington, DC: Peterson Institute for International Economics.

Gilley, B. (2000) 'Huawei's Fixed Line to Beijing', *Far Eastern Economic Review*, 28 December, pp. 94–98.

Goldin, C. and Katz, L. F. (2007) *The Race between Education and Technology: The Evolution of U.S. Educational Wage Differentials, 1890 to 2005.* NBER Working Paper No. 12984. Cambridge, MA: National Bureau of Economic Research.

Goldman, E. and Viswanath, P. V. (2017) 'Internal Capital Markets, Forms of Intragroup Transfers, and Dividend Policy: Evidence from Indian Corporates', *Journal of Financial Research*, 40(4), pp. 567–610.

Gomez, E. T. and Jomo K. S. (1998) *Malaysia's Political Economy: Politics, Patronage, and Profits*. 2nd ed. Cambridge, UK: Cambridge University Press.

Gompers, P. and Lerner, J. (2001) 'The Venture Capital Revolution', *Journal of Economic Perspectives*, 15(2), pp. 145–68.

Gopalan, R., Nanda, V. and Seru, A. (2007) 'Affiliated Firms and Financial Support: Evidence from Indian Business Groups', *Journal of Financial Economics*, 86(3), pp. 759–95.

Granovetter, M. (1973) 'The Strength of Weak Ties', *American Journal of Sociology*, 78(6), pp. 1360–80.

Granovetter, M. (1994) 'Business Groups', in Smelser, N. J. and Swedberg, R. (eds) *The Handbook of Economic Sociology*. Princeton, NJ: Princeton University Press, pp. 453–75.

Guo, L. (2014) 'Chinese Style VIEs: Continuing to Sneak under Smog', *Cornell International Law Journal*, 47(3), p. 569.

Hamdani, A., Kosenko, K. and Yafeh, Y. (2020) *Regulatory Measures to Dismantle Pyramidal Business Groups: Evidence from the United States, Japan, Korea and Israel*. CEPR Discussion Paper No. DP15342.

Hamilton, G. G. and Kao, C. (2018) *Making Money: How Taiwanese Industrialists Embraced the Global Economy*. Stanford, CA: Stanford University Press.

Harwit, E. (2017) 'WeChat: Social and Political Development of China's Dominant Messaging App', *Chinese Journal of Communication*, 10(3), pp. 312–27.

Hofstede, G. H. (1984) *Culture's Consequences: International Differences in Work-Related Values*. London: Sage Publications.

Hsieh, C.-T. and Klenow, P. J. (2009) 'Misallocation and Manufacturing TFP in China and India', *The Quarterly Journal of Economics*, 124(4), pp. 1403–48.

Huneeus, F. and Kim, I. S. (2020) *The Effects of Firms' Lobbying on Resource Misallocation*. Working Paper No. 2018-23.

Hussain, S. and Safdar, N. (2018) 'Tunneling: Evidence from Family Business Groups of Pakistan', *Business & Economic Review*, 10(2), pp. 97–122.

ILO (2018) *Women and Men in the Informal Economy: A Statistical Picture*. Geneva: ILO.

Iversen, T. and Soskice, D. W. (2019) *Democracy and Prosperity: Reinventing Capitalism Through a Turbulent Century*. Princeton, NJ: Princeton University Press.

Jack, W. and Suri, T. (2014) 'Risk Sharing and Transactions Costs: Evidence from Kenya's Mobile Money Revolution', *American Economic Review*, 104(1), pp. 183–223.

Jacobs, J. (1961) *The Death and Life of Great American Cities*. New York: Random House.

Javorcik, B., Fitriani, F., Iacovone, L., Varela, G. and Duggan, V. (2012) *Productivity Performance in Indonesia's Manufacturing Sector*. Jakarta: World Bank.

Jia, L. and Winseck, D. (2018) 'The Political Economy of Chinese Internet Companies: Financialization, Concentration, and Capitalization', *International Communication Gazette*, 80(1), pp. 30–59.

Jin, D. Y. (2017) 'Evolution of Korea's Mobile Technologies: A Historical Approach', *Mobile Media & Communication*, 6(1), pp. 71–87.

Johnson, S. and Mitton, T. (2003) 'Cronyism and Capital Controls: Evidence from Malaysia', *Journal of Financial Economics*, 67(2).

Kara, S. (2019) *Tainted Garments: The Exploitation of Women and Girls in India's Home-Based Garment Sector*. Berkeley, CA: Blum Center for Developing Economies, University of California.

Kaufmann, D. and Kraay, A. (2020) 'Worldwide Governance Indicators'. World Bank. Available at: info.worldbank.org/governance/wgi/#home

Keim, G. D. and Hillman, A. J. (2008) 'Political Environments and Business Strategy: Implications for Managers', *Business Horizons*, 51(1), pp. 47–53.

Keister, L. A. (1998) 'Engineering Growth: Business Group Structure and Firm Performance in China's Transition Economy', *American Journal of Sociology*, 104(2), pp. 404–40.

Khanna, T. and Palepu, K. (2000a) 'Is Group Affiliation Profitable in Emerging Markets? An Analysis of Diversified Indian Business Groups', *The Journal of Finance*, 55(2), pp. 867–91.

Khanna, T. and Palepu, K. (2000b) 'The Future of Business Groups in Emerging Markets: Long-Run Evidence from Chile', *The Academy of Management Journal*, 43(3), pp. 268–85.

Khanna, T. and Palepu, K. G. (2010) *Winning in Emerging Markets: A Road Map for Strategy and Execution*. Boston, MA: Harvard Business Press.

Khanna, T. and Rivkin, J. W. (2001) 'Estimating the Performance Effects of Business Groups in Emerging Markets', *Strategic Management Journal*, 22(1), pp. 45–74.

Khanna, T. and Yafeh, Y. (2007) 'Business Groups in Emerging Markets: Paragons or Parasites?', *Journal of Economic Literature*, 45(2), pp. 331–72.

Khorana, A., Shivdasani, A., Stendevad, C. and Sanzhar, S. (2011) 'Spin-offs: Tackling the Conglomerate Discount', *Journal of Applied Corporate Finance*, 23(4), pp. 90–101.

Kim, B., Pae, J. and Yoo, C.-Y. (2019) 'Business Groups and Tunneling: Evidence from Corporate Charitable Contributions by Korean Companies', *Journal of Business Ethics*, 154(3), pp. 643–66.

Kim, H. and Chung, K. H. (2018) 'Can State-Owned Holding (SOH) Companies Improve SOE Performance in Asia? Evidence from Singapore, Malaysia and China', *Journal of Asian Public Policy*, 11(2), pp. 206–25.

Klein, M. C. and Pettis, M. (2020) *Trade Wars Are Class Wars: How Rising Inequality Distorts the Global Economy and Threatens International Peace*. New Haven, CT: Yale University Press.

Kohli, H. S., Sharma, A. and Sood, A. (eds) (2011) *Asia 2050: Realizing the Asian Century*. 1st ed. Los Angeles, CA: SAGE Publications.

Komera, S., Lukose, P. J. J. and Sasidharan, S. (2018) 'Does Business Group Affiliation Encourage R&D Activities? Evidence from India', *Asia Pacific Journal of Management*, 35(4), pp. 887–917.

Kuznets, P. W. (1988) 'An East Asian Model of Economic Development: Japan, Taiwan, and South Korea', *Economic Development and Cultural Change*, 36(3), pp. S11–43.

Lamin, A. (2013) 'Business Groups as Information Resource: An Investigation of Business Group Affiliation in the Indian Software Services Industry', *Academy of Management Journal*, 56(5), pp. 1487–509.

Lane, N. (2021) *Manufacturing Revolutions: Industrial Policy and Industrialization in South Korea*. CSAE Working Paper. Oxford, UK: Oxford University.

Ledeneva, A. V. (1998) *Russia's Economy of Favours: Blat, Networking, and Informal Exchange*. Cambridge, UK/New York: Cambridge University Press (Cambridge Russian, Soviet and post-Soviet studies, 102).

Lee, Y. and Lee, K. T. (2015) 'Economic Nationalism and Globalization in South Korea: A Critical Insight', *Asian Perspective*, 39(1), pp. 125–51.

Leff, N. H. (1978) 'Industrial Organization and Entrepreneurship in the Developing Countries: The Economic Groups', *Economic Development and Cultural Change*, 26(4), pp. 661–75.

Levinson, M. (2016) *The Box: How the Shipping Container Made the World Smaller and the World Economy Bigger*. 2nd ed. Princeton, NJ: Princeton University Press.

Lewin, A. Y., Kenney, M. and Murmann, J. P. (2016) *China's Innovation Challenge: Overcoming the Middle-Income Trap*. Cambridge, UK: Cambridge University Press.

Liang, Z., Appleton, S. and Song, L. (2016) *Informal Employment in China: Trends, Patterns and Determinants of Entry*. IZA Discussion Paper 10139. Bonn: Institute of Labor Economics (IZA).

Lin, L. (2017) 'Venture Capital Exits and the Structure of Stock Markets in China', *Asian Journal of Comparative Law*, 12(1), pp. 1–40.

Lin, K. J., Lu, X., Zhang, J. and Zheng, Y. (2020) 'State-Owned Enterprises in China: A Review of Forty years of Research and Practice', *China Journal of Accounting Research*, 13(1), pp. 31–55.

Lins, K. and Servaes, H. (1999) 'International Evidence on the Value of Corporate Diversification', *Journal of Finance*, 54(6), pp. 2215–39.

Lins, K. V. and Servaes, H. (2002) 'Is Corporate Diversification Beneficial in Emerging Markets?', *Financial Management*, 31(2), pp. 5–31.

Lubenow, G. C. (1985) 'Jobs Talks About His Rise and Fall', *Newsweek*, 29 September. Available at: www.newsweek.com/jobs-talks-about-his-rise-and-fall-207016

Maddison, A. (2007) *Chinese Economic Performance in the Long Run, 960–2030 AD*. 2nd ed. Paris: Development Centre Studies, OECD Publishing.

Maddison, A. (2010) 'Statistics on World Population, GDP and Per Capita GDP, 1–2008 AD'. Available at: www.ggdc.net/maddison/oriindex.htm

Marshall, M. G., Gurr, T. R. and Jaggers, K. (2019) 'Polity IV Project: Political Regime Characteristics and Transitions, 1800–2018'. Center for Systemic Peace. Available at: www.systemicpeace.org/inscr/

Mason, A. D. and Shetty, S. (2019) *A Resurgent East Asia: Navigating a Changing World*. World Bank East Asia and Pacific Regional Report. Washington, DC: World Bank.

Mason, C. and Brown, R. (2014) *Entrepreneurial Ecosystems and Growth Oriented Entrepreneurship*. Background Paper Prepared for the Workshop Organised by the OECD LEED Programme and the Dutch Ministry of Economic Affairs. OECD.

Mazzucato, M. (2015) *The Entrepreneurial State: Debunking Public vs. Private Sector Myths*. Rev. ed. New York: Public Affairs.

McDonald, H. (2010) *Ambani & Sons*. New Delhi: Roli Books.

McGuire, J. and Dow, S. (2003) 'The Persistence and Implications of Japanese Keiretsu Organization', *Journal of International Business Studies*, 34(4), pp. 374–388.

McKinsey Global Institute (2017) *A Future That Works: Automation, Employment, and Productivity*. London: McKinsey Global Institute.

Medina, L. and Schneider, F. (2018) *Shadow Economies Around the World: What Did We Learn Over the Last Twenty Years?* IMF Working Paper 18/17. Washington, DC: International Monetary Fund.

Megginson, W. L. (2016) 'Privatization, State Capitalism, and State Ownership of Business in the Twenty-first Century', *Foundations and Trends in Finance*. 11 (1–2), pp. 1–153.

Megginson, W. L. (2017) 'Privatization Trends and Major Deals in 2015 and 2016'. Available at: https://papers.ssrn.com/sol3/papers.cfm?abstract_id=2944287

Milanović, B. (2016) *Global Inequality: A New Approach for the Age of Globalization*. Cambridge, MA: The Belknap Press of Harvard University Press.

Milanović, B. (2019) *Capitalism, Alone: The Future of the System That Rules the World*. 1st ed. Cambridge, MA: The Belknap Press of Harvard University Press.

Mokyr, J. (1990) *The Lever of Riches: Technological Creativity and Economic Progress*. New York: Oxford University Press.

Morck, R. (2005) 'How to Eliminate Pyramidal Business Groups: The Double Taxation of Intercorporate Dividends and Other Incisive Uses of Tax Policy', in Poterba, J. M., *Tax Policy and the Economy*. Cambridge, MA: MIT Press, pp. 135–79.

Morck, R., Wolfenzon, D. and Yeung, B. (2005) 'Corporate Governance, Economic Entrenchment, and Growth', *Journal of Economic Literature*, 43(3), pp. 655–720.

Musacchio Farias, A. and Lazzarini, S. G. (2014) *Reinventing State Capitalism: Leviathan in Business, Brazil and Beyond*. Cambridge, MA: Harvard University Press.

Myrdal, G. (1972) *Asian Drama; An Inquiry into the Poverty of Nations*. 1st ed. New York: Pantheon Books.

Nahid, F., Gomez, E. T. and Yacob, S. (2019) 'Entrepreneurship, State–Business Ties and Business Groups in Bangladesh', *Journal of South Asian Development*, 14 (3), pp. 367–90.

NASSCOM (2019) *18% of The Total Start-Ups in 2019 Are Now Leveraging Deep Tech – NASSCOM Start-Up Report 2019*. Bengaluru: NASSCOM. Available at: nasscom.in/sites/default/files/media_pdf/pr-indian_start_up%20_ecosystem .pdf

Natural Earth (2021) '1.50m Cultural Vectors: Admin 0-Countries'. Available at: www.naturalearthdata.com/downloads/50m-cultural-vectors/

Nature (2020) 'Nature Index'. Available at: www.natureindex.com

Needham, J. (1954–2015) *Science and Civilisation in China*. Cambridge, UK: Cambridge University Press.

Nikkei Asia (2016) 'Indonesian Media Tycoon Juggles Business and Politics', 3 March. Available at: https://asia.nikkei.com/Business/Indonesian-media-tycoon-Hary-Tanoe-juggles-business-and-politics

Nikkei Asia (2019a) 'China Overtakes US in AI Patent Rankings', 10 March. Available at: asia.nikkei.com/Business/Business-trends/China-overtakes-US-in-AI-patent-rankings

Nikkei Asia (2019b) 'Crony Capital: How Duterte Embraced the Oligarchs', 4 December. Available at: asia.nikkei.com/Spotlight/The-Big-Story/Crony-cap ital-How-Duterte-embraced-the-oligarchs

Nikkei Asia (2019c) 'The Chinese-Indonesian Dynasty Reshaping a $30bn Empire via WhatsApp', 27 December. Available at: asia.nikkei.com/Spotlight/Asian-Family-Conglomerates/The-Chinese-Indonesian-dynasty-reshaping-a-30bn-empire-via-WhatsApp

Nolan, P. (2001) *China and the Global Economy: National Champions, Industrial Policy and the Big Business Revolution*. London, UK: Palgrave Macmillan UK.

North, D. C. (1990) *Institutions, Institutional Change and Economic Performance*. 1st ed. Cambridge: Cambridge University Press.

North, D. C. (1991) 'Institutions', *Journal of Economic Perspectives*, 5(1), pp. 97–112.

Numazaki, I. (1993) 'The Tainanbang: The Rise and Growth of a Banana-Bunch-Shaped Business Group in Taiwan', *The Developing Economies*, 31(4), pp. 485–510.

O'Connor, A., Stam, E., Sussan, F. and Audretsch, D. B. (eds) (2017) *Entrepreneurial Ecosystems: Place-Based Transformations and Transitions*. New York: Springer.

OECD (1997) *National Innovation Systems*. Paris: OECD Publishing.

OECD (2011) *Reform Priorities in Asia: Taking Corporate Governance to a Higher Level*. Paris: OECD Publishing.

OECD (2017) *OECD Digital Economy Outlook 2017: Spotlight on Korea*. Paris: OECD Publishing.

OECD (2018a) *Competition Law in Asia-Pacific: A Guide to Selected Jurisdictions*. Paris: OECD Publishing.

OECD (2018b) *Entrepreneurship at a Glance: 2018 Highlights*. Paris: OECD Publishing.

OECD (2018c) 'Strengthening SMEs and Entrepreneurship for Productivity and Inclusive Growth', *2018 OECD Ministerial Conference on SME*. Paris: OECD Publishing.

OECD (2020) *PISA 2018 Database*. Paris: OECD Publishing. Available at: www .oecd.org/pisa/data/2018database/

OECD and ADB (2019) *Government at a Glance Southeast Asia 2019*. Paris: OECD Publishing.

OECD, DynEmp (2016) *No Country for Young Firms?* Directorate for Science, Technology and Innovation Policy Note.

Ofer, G. (1987) 'Soviet Economic Growth: 1928–1985', *Journal of Economic Literature*, 25(4), pp. 1767–833.

Oh, J. and Zolt, E. M. (2018) 'Wealth Tax Add-Ons: An Alternative to Comprehensive Wealth Taxes', *UCLA School of Law, Law-Econ Research Paper*, 18(3). Available at: https://papers.ssrn.com/sol3/papers.cfm?abstract_id=3167483

One Road Research (2018) *Asia's Powerful Conglomerates*. Research Note. Hong Kong.

Ortiz-Ospina, E. and Roser, M. (2017) 'Happiness and Life Satisfaction'. Available at: ourworldindata.org/happiness-and-life-satisfaction

Pakistan Enquiry Committee Report on Sugar Prices (2020) *Report of the Commission of Inquiry Constituted by Ministry of Interior to Probe into the Increase in Sugar Prices*, Islamabad.

Panagariya, A. (2008) *India: The Emerging Giant*. Oxford, UK/New York: Oxford University Press.

Park, A., Wu, Y. and Du, Y. (2012) *Informal Employment in Urban China: Measurement and Implications*. Washington, DC: World Bank.

Pattnaik, C., Lu, Q. and Gaur, A. S. (2018) 'Group Affiliation and Entry Barriers: The Dark Side Of Business Groups in Emerging Markets', *Journal of Business Ethics*, 153(4), pp. 1051–66.

PCC (2020) 'PCC Issues Rules on Merger Reviews Under Bayanihan 2'. Available at: www.phcc.gov.ph/press-releases/pcc-mao-rules-bayanihan2/

Pei, M. (2016) *China's Crony Capitalism: The Dynamics of Regime Decay.* Cambridge, MA: Harvard University Press.

Philippon, T. (2019) *The Great Reversal: How America Gave Up on Free Markets.* Cambridge, MA: The Belknap Press of Harvard University Press.

Piketty, T. (2014) *Capital in the Twenty-First Century.* Translated by A. Goldhammer. Cambridge, MA: The Belknap Press of Harvard University Press.

Piketty, T., Yang, L. and Zucman, G. (2019) 'Capital Accumulation, Private Property, and Rising Inequality in China, 1978–2015', *American Economic Review,* 109(7), pp. 246996.

PopulationPyramid.net (2020) 'Population Pyramids of the World from 1950 to 2100'. PopulationPyramid.net. Available at: www.populationpyramid.net/

Porter, M. E. (1990) *The Competitive Advantage of Nations.* 1st ed. New York: Free Press.

Poutziouris, P., Savva, C. S. and Hadjielias, E. (2015) 'Family Involvement and Firm Performance: Evidence from UK Listed Firms', *Journal of Family Business Strategy,* 6(1), pp. 14–32.

Pritchett, L. and Summers, L. (2013) 'Asiaphoria Meets Regression to the Mean', in Glick, R. and Spiegel, M. (eds) *Prospects for Asia and the Global Economy. Asia Economic Policy Conference,* San Francisco, CA: Federal Reserve Bank of San Francisco, pp. 33–71.

Puri, S. (2020) 'Research: How Corporate Boards in Asia Can Improve Governance', *Harvard Business Review,* 13 July. Available at: https://hbr.org/2020/07/research-how-corporate-boards-in-asia-can-improve-governance

Quartz (2019) *The "SoftBank of China" Has Invested in More Unicorns Than SoftBank, Quartz.* Available at: qz.com/1733132/chinas-tencent-has-invested-in-more-unicorns-than-softbank/

Rahman, M., Bhattacharya, D. and Hasan, Md. Al. (2018) *The Role of the Informal Sector in Inclusive Growth: A State of Knowledge Study from Policy Perspectives.* San Francisco: The Asia Foundation.

Rajan, R. and Zingales, L. (2005) 'Which Capitalism? Lessons from the East Asian Crisis', *Journal of Applied Corporate Finance,* 17(2), pp. 40–48.

Rajshekhar, M. (2019a) *From 2014 to 2019: How the Adani Group Funded Its Expansion, Scroll.in.* Available at: scroll.in/article/923201/from-2014-to-2019-how-the-adani-group-funded-its-expansion

Rajshekhar, M. (2019b) *From 2014 to 2019: How the Adani Group's Footprint Expanded Across India, Scroll.in.* Available at: scroll.in/article/923095/from-2014-to-2019-how-the-adani-groups-footprint-expanded-across-india

Report of the Inquiry Committee Constituted by the Prime Minister of Pakistan Regarding Increase in Sugar Prices (2020) Islamabad, Pakistan. 24 March.

Available at: www.app.com.pk/pdf/Commission-Reports/Sugar/Sugar-Inquiry-Committee-Main-Report-dated-24.03.2020.pdf

Ridley, M. (2020) *How Innovation Works: And Why It Flourishes in Freedom*. London: HarperCollins.

Rodrik, D. (2009) 'Industrial Policy: Don't Ask Why, Ask How', *Middle East Development Journal*, 1(1), pp. 1–29.

Rodrik, D. (2015) *Premature Deindustrialization*. NBER Working Paper No. 20935. Cambridge, MA: National Bureau of Economic Research.

Rodrik, D. (2018) *New Technologies, Global Value Chains, and Developing Economies*. NBER Working Paper No. 25164. Cambridge, MA: National Bureau of Economic Research. DOI: http://doi.org/10.3386/w25164

Ruangkanjanases, A., Posinsomwong, N. and Chen, C. (2015) 'The Application of Confucius Practice in Management at the Largest Agriculture-Based Conglomerate Group of Companies in Thailand', *International Journal of Social Science and Humanity*, 4, pp. 354–61.

Rumelt, R. P. (1974) *Strategy, Structure, and Economic Performance*. Boston, MA: Division of Research, Grad. School of Business Administration, Harvard University.

Saxenian, A. (1994) *Regional Advantage: Culture and Competition in Silicon Valley and Route 128*. Cambridge, MA: Harvard University Press.

Saxenian, A. and Quan, X. I. (2005) 'Guanxi and Government: The Chinese Software Industry in Transition', in Commander, S. (ed.) *The Software Industry in Emerging Markets*. Cheltenham, UK/ Northampton, MA: Edward Elgar.

Schipke, A. (2015) *Frontier and Developing Asia: The Next Generation of Emerging Markets, Frontier and Developing Asia*. Washington, DC: International Monetary Fund.

Schoff, J. L. and Ito, A. (2019) *Competing with China on Technology and Innovation*. Alliance Policy Coordination Brief. Washington, DC: Carnegie Endowment for International Peace. Available at: carnegieendowment.org/2019/10/10/competing-with-china-on-technology-and-innovation-pub-80010

Schumpeter, J. A. (1934) *The Theory of Economic Development*. Translated by R. Opie. Cambridge, MA: Harvard University Press.

Schumpeter, J. A. (1942) *Capitalism, Socialism and Democracy*. New York: Harper & Brothers.

Schwab, K. (2016) *The Fourth Industrial Revolution*. 1st ed. New York: Crown Business.

Shen, H. (2019) 'China's Tech-Giants: Baidu, Alibaba, Tencent', in *Panorama Insights into Asian and European Affairs: Digital Asia*. Singapore: Konrad-Adenauer-Stiftung, pp. 33–41.

Shum, D. (2021) *Red Roulette: An Insider's Story of Wealth, Power, Corruption and Vengeance in Today's China*. London: Simon & Schuster.

Siegel, J. and Choudhury, P. (2012) 'A Reexamination of Tunneling and Business Groups: New Data and New Methods', *The Review of Financial Studies*, 25(6), pp. 1763–98.

Solow, R. M. (1957) 'Technical Change and the Aggregate Production Function', *The Review of Economics and Statistics*, 39(3), pp. 312–20.

Spalding, R. S. (2019) *Stealth War: How China Took Over While America's Elite Slept*. New York: Portfolio/Penguin.

Susskind, D. (2020) *A World Without Work: Technology, Automation, and How We Should Respond*. 1st ed. New York: Metropolitan Books/Henry Holt & Company.

Tang, M. (2020) 'From "Bringing-In" to "Going-Out": Transnationalizing China's Internet Capital Through State Policies', *Chinese Journal of Communication*, 13(1), pp. 27–46.

The Economist (2020a) 'A Slump Exposes Holes in China's Welfare State', 7 May. Available at: www.economist.com/china/2020/05/07/a-slump-exposes-holes-in-chinas-welfare-state

The Economist (2020b) 'Ant Group and Fintech Come of Age', 8 October. Available at: www.economist.com/leaders/2020/10/08/ant-group-and-fintech-come-of-age

The Edge (2019) 'Local Banks, Family-Owned Blue Chips Pay Good Dividends; Malaysian Families Reward Through CEO Pay, Directors' Fees', 24 June.

Tihanyi, L., Aguilera, R. V., Heugens, P., van Essen, M., Sauerwald, S., Duran, P. and Turturea, R. (2019) 'State Ownership and Political Connections', *Journal of Management*, 45(6), pp. 2293–321.

Tybout, J. R. (2000) 'Manufacturing Firms in Developing Countries: How Well Do They Do, and Why?', *Journal of Economic Literature*, 38(1), pp. 11–44.

UNCTAD (2020) 'Foreign Direct Investment: Inward and Outward Flows and Stock, Annual'. UNCTAD. Available at: unctadstat.unctad.org/EN/Index.html

United Nations Conference on Trade and Development (UNCTAD) (2019) *World Investment Report 2019: Special Economic Zones*. Available at: https://unctadstat.unctad.org

UNIDO (2020) *Industrial Development Report 2020. Industrializing in the Digital Age*. Vienna: UNIDO.

Villalonga, B. and Amit, R. (2006) 'How Do Family Ownership, Control and Management Affect Firm Value?', *Journal of Financial Economics*, 80(2), pp. 385–417.

Wall Street Journal (2019) 'State Support Helped Fuel Huawei's Global Rise', 25 December. Available at: www.wsj.com/articles/state-support-helped-fuel-huaweis-global-rise-11577280736?mod=article_inline

Wall Street Journal (2020) '"Samsung Rising" Review: The Republic of Samsung', 22 March. Available at: www.wsj.com/articles/samsung-rising-review-the-republic-of-samsung-11584906258

Wernerfelt, B. and Montgomery, C. A. (1988) 'Tobin's q and the Importance of Focus in Firm Performance', *The American Economic Review*, 78(1), pp. 246–50.

Williamson, O. E. (2000) 'The New Institutional Economics: Taking Stock, Looking Ahead', *Journal of Economic Literature*, 38(3), pp. 595–613.

Williamson, P. J. and Yin, E. (2014) 'Accelerated Innovation: The New Challenge From China', *MIT Sloan Management Review*, Magazine Summer, July. Available at: https://sloanreview.mit.edu/article/accelerated-innovation-the-new-challenge-from-china/

WIPO (2019) *World Intellectual Property Indicators 2019*. Geneva: World Intellectual Property Organization.

WIPO, Cornell University and INSEAD (2019) *The Global Innovation Index 2019: Creating Healthy Lives – The Future of Medical Innovation*. Geneva: World Intellectual Property Organization.

Witt, M. A. and Lewin, A. Y. (2007) 'Outward Foreign Direct Investment as Escape Response to Home Country Institutional Constraints', *Journal of International Business Studies*, 38(4), pp. 579–94.

Wittfogel, K. A. (1957) *Oriental Despotism: A Comparative Study of Total Power*. New Haven, CT: Yale University Press.

World Bank (2013) *World Development Report 2013: Jobs*. Washington, DC: World Bank.

World Bank (2017) *Trouble in the Making?: The Future of Manufacturing-Led Development*. Washington, DC: World Bank.

World Bank (2018a) *Growing Smarter: Learning and Equitable Development in East Asia and Pacific*. World Bank East Asia and Pacific Regional Report. Washington, DC: World Bank.

World Bank (2018b) *Poverty and Shared Prosperity 2018: Piecing Together the Poverty Puzzle*. Washington, DC: World Bank.

World Bank (2018c) *Riding the Wave: An East Asian Miracle for the 21st Century*. World Bank East Asia and Pacific Regional Report. Washington, DC: World Bank.

World Bank (2020a) *Atlas of Social Protection Indicators of Resilience and Equity (ASPIRE)*. Washington, DC: World Bank. Available at: www.worldbank.org/en/data/datatopics/aspire

World Bank (2020b) *Doing Business 2020: Comparing Business Regulation in 190 Economies*. Washington, DC: World Bank.

World Bank (2020c) *Enhancing Government Effectiveness and Transparency: The Fight Against Corruption*. Kuala Lumpur: World Bank.

World Bank (2020d) *Global Productivity: Trends, Drivers, and Policies*. Washington, DC: World Bank.

World Bank (2020e) *Making It Big: Why Developing Countries Need More Large Firms*. Washington, DC: World Bank.

World Bank (2020f) *World Development Indicators*. Washington, DC: World Bank. Available at: databank.worldbank.org/source/world-development-indicators

World Bank Group and Development Research Center of the State Council, P. R. China (2019) *Innovative China: New Drivers of Growth*. Washington, DC: World Bank.

'World Inequality Database' (2020). Available at: wid.world

WTO Statistical Review (2019) *World Trade Statistical Review 2019*. Geneva: WTO.

Yip, G. S. and McKern, B. (2016) *China's Next Strategic Advantage: From Imitation to Innovation*. Cambridge, MA: The MIT Press.

Young, A. (1995) 'The Tyranny of Numbers: Confronting the Statistical Realities of the East Asian Growth Experience', *The Quarterly Journal of Economics*, 110 (3), pp. 641–80.

Yu, F. L. T. (2018) 'Private Enterprise Development in a One-Party Autocratic State: The Case of Alibaba Group in China's E-Commerce', *Issues & Studies*, 54(1), pp. 1–33.

Zhang, C. (2019) *How Much Do State-Owned Enterprises Contribute to China's GDP and Employment?* Washington, DC: World Bank.

Index

Printed by Printforce, United Kingdom